THE METAPHYSICS OF
VIRTUAL REALITY

Other Books by the Author

Electric Language: A Philosophical Study of Word Processing (1987)

Translation of Martin Heidegger, *The Metaphysical Foundations of Logic* (1984)

THE METAPHYSICS OF VIRTUAL REALITY

Michael Heim

New York | Oxford | OXFORD UNIVERSITY PRESS | 1993

Oxford University Press

Oxford   New York   Toronto
Delhi   Bombay   Calcutta   Madras   Karachi
Kuala Lumpur   Singapore   Hong Kong   Tokyo
Nairobi   Dar es Salaam   Cape Town
Melbourne   Auckland   Madrid

and associated companies in
Berlin   Ibadan

Copyright © 1993 by Michael Heim

Published by Oxford University Press, Inc.
200 Madison Avenue, New York, New York 10016

Oxford is a registered trademark of Oxford University Press, Inc.

Library of Congress Cataloging-in-Publication Data
Heim, Michael, 1944–
The metaphysics of virtual reality / by Michael Heim.
p. cm.   Includes bibliographical references and index.
ISBN 0-19-508178-1
1. Human–computer interaction. 2. Virtual reality.
3. Technology—Social aspects. I. Title.
QA76.9.H85H45 1993
006—dc20

9 8 7 6 5 4 3 2 1

Printed in the United States of America
on acid-free paper

For young Mike,
Jack and Dorothy,
and always
Joanna

In the present state of the world, the control we have of physical energies, heat, light, electricity, etc., without control over the use of ourselves is a perilous affair. Without control of ourselves, our use of other things is blind.

John Dewey,
Preface to F. M. Alexander,
The Resurrection of the Body

FOREWORD

If virtual reality were just another technology, you would not
have heard so much about it. However, it is a technology that
can be applied to every human activity and can be used to
mediate in every human transaction. Since you are completely
immersed in the virtual world, virtual reality constitutes a
new form of human experience—one that may be as important
to the future as film, theater, and literature have been to the
past. Its potential impact is so broad that it may define the culture that results from its use. As a consequence, the concept
of virtual reality will be as widely used as a metaphor as it is
in practice.

In this book, Michael Heim points out that virtual reality
is the culmination of a process that has been going on for
some time in technology and for even longer in Western
thought. He goes back to the roots of this development and
traces how each change in the technology of knowledge has
led to a change in our relationship to knowledge and ultimately in our view of ourselves. His review of the ideas of
familiar philosophers in light of contemporary technological
change is fascinating. For instance, Leibniz's vision of a community of minds precisely anticipates current data networks.

The beginning of this evolution occurred when the invention of writing enormously expanded the capacity of human
memory to extend beyond a single lifetime. But until the invention of the printing press, writing was itself a form of wor-
ship. After Gutenberg, the written word could be owned by
individuals as well as institutions.

Recently, such is the pace of development and so rapid

is our adaptation to it that one revolution is not over before the next has begun. One wonders if we are in the midst of a cataclysm that will soon run its course, or if the fact not just of change but of accelerating change has now become a constant feature of our lives.

In the past two decades, the printed word has begun to yield to the electronic word. Whereas the printing press minimized the reader's investment, word processing reduces the writer's. Writers can write to see if they have anything to say. If a thought emerges, they can capture it, tame it, and make it march on parade as though it had been conceived to do so. Recently, new tools for thought processing have appeared to further augment human thinking.

While word processing expedites the traditional writing task, hypertext breaks with the linear sequence of ordered thought demanded by the printed word. The reader is asked to make decisions about order of presentation that the author was not willing to make.

The nonlinear, free-association format of hypertext has been incorporated into multimedia—the most recent revolution. In this new technology, the black-and-white, static, symbolic, sensory-deprivation world of the intellect has begun to yield to multisensory modes of presentation. The awkward juxtaposition of pictures and words that exists in print gives way to a new form of expression in which illustrations dominate. By totally immersing the consumer in illustration, virtual reality is the momentary culmination of this evolution.

However, virtual reality changes our relationship to information in an even more fundamental way. It is the first intellectual technology that permits the active use of the body in the search for knowledge. Does this imply the death of abstract symbols and with it the demise of the sedentary intellect? Or will we invent new three-dimensional, colored, animated symbols that will interact with us rather than waiting passively for us to read them? Will this rejoining of the mind and the body create a new breed of intellectual?

Virtual reality raises many philosophical questions. What is the status of virtual experience? Although we typically make a clear distinction between fact and fiction, we are increasingly the products of artificial experience. This can be good for pilots who train from carefully composed scenarios that distill the cumulative flight experience of all humankind. But what of a world in which every action is rehearsed in simulation before it is taken, as was the case for pilots during the Gulf war? Will real action lose its immediacy when it is but a recapitulation of simulated activity?

In virtual reality, traditional philosophical questions are no longer hypothetical. What is existence? How do we know? What is reality? Who am I? These are aesthetic issues with engineering consequences. They are certainly not remote or esoteric, given the possibility of creating artificial experiences that are as compelling as the real ones.

In this book, Heim puts his finger on why virtual reality has excited us as a culture. It is an intellectual feast to which we are all invited, an intellectual frontier that we are all free to explore and invent. He analyzes this process of change without judging whether it is good or bad, because understanding it is both important and interesting. We can embrace it not because it will cost us nothing, but because it is our destiny to redefine ourselves. In the world of the book, the truth is an immutable thing to be captured and recorded. But as the age of electronic information matures and the age of artificial experience commences, we recognize that we are on a journey and while we may question whether the next destination is better than the place we left, we realize that the journey is ours, for we must see what it is—that what we have made, makes us.

Myron E. Krueger

PREFACE

As these words spread across my computer screen, I occasionally look up at a sublime sight, the high peaks of the Canadian Rocky Mountains. Every few minutes, the wind
whistles through the sharp, snow-covered peaks, and the si-
lence beneath these November winds runs so deep that it
catches your breath and holds you tight in a deep stillness.
You realize then why this spot became a famous gathering
place for Canada's artists, the Banff Centre for the Arts in Al-
berta. Up here at 4,500 feet, you see big elk grazing quietly
at the roadside. But a paradox disrupts the scenic beauty. The
artists and writers in residence at this haven for the arts have
come here to plan and construct virtual-reality prototypes,
computer-simulated environments, and artificial worlds.
Plugged into electric power and computer chips, the human
race in this last decade of the twentieth century is preparing
to lift off from nature into another—electronic—space.

In a previous book, *Electric Language*, I used the computer as a seismograph for measuring ontological shifts, the
changes in our contemporary reality. Seen philosophically, the
word processor creates a new relationship to symbols, to language, and, by extension, to reality. One aspect of the new reality is a powerful feedback mechanism that now undergirds
our culture. The media today draw on worldwide computer
links, speeding up communications by radio, newspapers, and
television. Computer networks have sprung up to form a giant
web for human exchange. Ideas fly back and forth, circling the
globe at the speed of electricity. This new electric language

forms an instant feedback loop, the likes of which have never before existed. And not a moment too soon.

The information infrastructure comes just in time. It brings a cybernetic dimension that allows us to join together to discuss and critique imminent technologies. Major technological breakthroughs loom on the horizon, such as the advent of virtual reality, the totally immersive computer simulation. Previously, a technology would be germinated by inventors, be tested by engineers, be produced by developers, and then be sent to market to transform a culture. But, the universal computer network means that no technological breakthrough comes to market without first passing over the platform of public appraisal. The electronic net captures everything, first as a shared project, then as news, and finally as an issue for debate. The network fosters ongoing discussion in which interaction runs at high speed and one person can connect to many, bypassing the established hierarchies. While such openness may hassle the technically minded, it inserts a new dimension into the world and establishes a new relationship to technology, a symbiosis in which neither the human nor the machine dominates. Just as the cybernaut can gesture to create or alter objects in a virtual environment, so too the human race now inhabits a world in which almost everything we recognize results from our own doing.

Virtual reality (VR) was born in this cybernetic dimension. From its embryonic stage, VR has been under constant social observation and discussion. Not left to technicians, VR remains a focal issue of interdisciplinary and lay debate— even before it has fully emerged from the cocoon of research and development. Obviously, as we convert to VR, the net too will undergo a change as we move in and out of a self-subsistent computer-generated world. But the question will still be pressing. How much can humans change and still remain human as they enter the cyberspace of computerized realities?

That question shattered my previous reflections on the computer rather abruptly in 1989 when I first entered VR. Instead of sitting before a screen with keyboard or mouse, I donned a helmet and glove and felt immersed in a computer-generated environment. No longer outside the computer, I walked through the looking glass. My philosophical seismograph went crazy. This virtual reality I tried was still a primitive prototype, like arcade games and amateur flight simulators. But its implications seemed enormous. The shifts in reality I had found in earlier computer use were subtle by comparison. The ontological shift through digital symbols became in VR a full-fledged, aggressive, surrogate reality.

This book of essays continues the line of thought in my earlier book. Only this time, things are more urgent, and the question of the ontological shift—the reality shift—has become more manifest.

When I write of an ontological shift, I mean more than a change in how we humans see things, more than a paradigm shift or a switch in our epistemological stance. Of course, our access to knowledge changes dramatically as we computerize the arts, sciences, and business. But there are more things in heaven and earth than are dreamed of in our epistemology. An ontological shift is a change in the world under our feet, in the whole context in which our knowledge and awareness are rooted. Things change even before we become aware of what has been happening to us. We might look upon the automobile, for instance, as a limited tool, as a human device for transportation. In fact, however, the world itself changed when we introduced the automobile. The widespread use of automobiles opens us up to different places, and these places connect in new ways that differ from the old places in kind and quality, and the world we live in changes gradually but inevitably. Knowledge in a scientific sense can lag only slightly behind this world transformation because knowledge becomes transformed in the process. The holistic background or world is the basic reality underlying our knowledge and awareness. Ontology, the study of being, is the effort to develop a peripheral vision by which we perceive and articulate the hidden background of beings, the world or context in which they become real and meaningful.

The chapters of this book mirror the progression from digital to virtual reality. The first several chapters offer evidence of an ontological shift, looking at how our daily activities on digital computers shape our reading and searching through information. Hypertexts and outliners, electronic mail and database searching, all belong to the new way things are organized. In my analysis, I try to show how this digital symbolic world brings both gains and losses. Each chapter usually offers some suggestions as to how we might preserve the better aspects of predigital reality in order to balance the technology that is changing our given reality.

Chapter 1, "Infomania": Over the ten years of the 1980s, a dramatic change took place in the production and storage of written language. An estimated 80 percent of written language began existing in digital form. Computers swallowed the cultural heritage of English-speaking countries. As the author of *Electric Language*, I was asked to summarize what I saw in the shift underlying this development. The result is "Infomania," which first appeared in Christopher Ricks and Leonard Michaels, eds., *The State of the Language* (Berkeley: University of California Press, 1990), and this article was later reprinted in the London newspaper *The Independent*. "Infomania" subsequently appeared in the international journal *Electric Word* in March 1990, in which I was the cover feature.

Chapter 2, "Logic and Intuition": *Electric Language* argued that our whole psychic framework changes as we use computers to read and write. Today 40 million computers in the United States alone spew out the findings of our infomania. Searching through the digital jungle alters the relationship of logic to intuition. While computers now handle the linear processes of creating, storing, and accessing data, the human being must still match the patterns and recognize the significance of these patterns. This essay traces the new division of tasks by recalling the age-old battle between logical and intuitive approaches. The central philosopher here is Blaise Pascal, who developed computer prototypes and who was himself something of a prototype for today's cybersage. "Logic and Intuition" was researched and written at the invitation of Harvey Wheeler, professor of information studies at the University of Southern California, for a session of EDU-

COM, though the paper was not presented, for complex technical reasons.

Chapter 3, "Hypertext Heaven": American audiences burst into applause when the *Millennium Falcon* (in the movie *Star Wars*) first jumped into hyperspace. Most computer users cheered, too, when they first discovered hypertext. Hypertext now appears everywhere on computers, from tax-preparation programs to encyclopedic guides to the Louvre. Like a joyride, hypertext has a kinesthetic appeal. But travel at such high speeds exacts a price. Just as the computer's search powers extend our knowledge while narrowing our focus, hypertext brings blissful browsing but weaves a dangerous illusion of God-like omniscience. "Hypertext Heaven" was part of my "Reflections on the Computer Screen," an electronic lecture (70 kilobytes) first given in February 1990 at the New School for Social Research graduate program in communications directed by Professor Paul Levinson; it also appeared on line in a different version in June 1990 in the electronic journal *PostModern Culture*, edited by John Unsworth and Elaine Orr at North Carolina State University, where it is still available in that version from pmc@ncsuvm.bitnet or jmueg@ncsuvm.bitnet.

Chapter 4, "Thought Processing": Software has a hidden agenda. As we computerize, we go through a lengthy learning process during which we enjoy few benefits and undergo much pain. Soon we adapt and become productive again, and the software seems transparent to the tasks we face. Yet when we take a closer look at one specific computer application, the outliner, we discover how software restructures our thought process. Throughout history, the written outline has changed its users and has even served as the tool for educational reformers who wished to—and did—transform society. Today's computerized outliners reveal the hidden agenda underlying software of all kinds. "Thought Processing" appears for the first time here.

Chapter 5, "Heidegger and McLuhan: The Computer as Component": The two names connect on the page like the meeting of Godzilla and King Kong. Widely known philosophical names of the twentieth century, familiar in both Europe and the United States, belong to two intellectual giants who saw technology as the central issue of the twentieth century: Heidegger giving technology a reality status and McLuhan discovering that no meaning escapes the mesh of the electronic media. Future scholars will sort out how these two thinkers differ while sharing many assumptions. More important to us is what we can learn from them about the role computers play in our lives. Both Heidegger and McLuhan saw that the computer would pose less danger to us as a rival artificial intelligence than it would as an intimate component of our everyday thought and work. I wrote about this for the journal *Philosophy and Literature,* published by Johns Hopkins University Press, in which a different version of this essay appeared in October 1992.

The later chapters of this book explore the shift of ground, the move from digital to virtual reality.

Chapter 6, "From Interface to Cyberspace": The last twenty years of amazing technological progress were driven by a definite logic, which followed the expansion and penetration of the interface by which humans interact with technical systems: machines became appliances, appliances offered an interface, the interface opened to cyberspace, and cyberspace offered virtual worlds to explore. With each step in the progression came a corresponding response by the public who implemented and used these technical systems. A logic of feeling underlies the public reactions to the technical development, "From Interface to Cyberspace" appears here for the first time and marks the shift of theory from digital to virtual reality. This 1992 essay traces the two sides of a coin: the logic of technical development and the logic of feeling in the human users.

Chapter 7, "The Erotic Ontology of Cyberspace": William Gibson's fiction first brought the term *cyberspace* into use among imaginative, technical people. His fiction imagines and magnifies the trade-offs in the move to virtual worlds. "The Erotic Ontology of Cyberspace" explored these trade-offs for the First Conference on Cyberspace, held at the University of Texas at Austin, May 4–5, 1990. The essay then appeared in Michael Benedikt, ed., *Cyberspace: First Steps* (Cambridge, Mass.: MIT Press, 1991). Sparks from this essay flew in many directions, one reaching Laurence Rozier, an expert on New Orleans voodoo, who builds and sells hypertexts that connect jazz, rock-and-roll, artificial intelligence, and African spiritual traditions.

Chapter 8, "The Essence of VR": In the late 1980s, the computer interface turned inside out and became a virtual world that people could enter. Jaron Lanier aptly termed the phenomenon *virtual reality*, although many noticed that the concept goes back to Myron Krueger in the 1960s and to Ivan Sutherland and Morton Heilig even earlier. Some observers, in their first efforts to explain the phenomenon to their contemporaries, pointed to drugs or sex or entertainment. But the profundity of the VR experience calls for something of a grander stature, something philosophical and religious. The time has come to grasp the phenomenon in its depth and scope. After all, we are talking about virtual "reality," not fleeting hallucinations or cheap thrills. We are talking about a profound shift in the layers of human life and thought. We are talking about something metaphysical.

Chapter 9, "Virtual-Reality Check": This is a revised version of a paper that appeared under the title "The Metaphysics of Virtual Reality" in the trade journal *Multimedia Review* in 1990, published by Meckler. It then appeared in a Meckler reprint called "Virtual Reality." The paper became the basis of a talk for the program "Virtual Reality: Theory, Practice, and Promise," held in San Francisco on December 10, 1990, which also became a videotape sold by Meckler.

Chapter 10, "The Electronic Café Lecture": Over a hundred people crowded into the Electronic Café International on Eighteenth Street in Santa Monica on March 12, 1992. They came from the computer industry, the entertainment field, and many other professions. They had come for the first meeting of the Southern California Virtual Reality Special Interest Group. The café, a model for the cyberarts since the 1984 Olympics, has become a center for experiments in telepoetry, electronic dance, cinema, and other multimedia experiments. The café's owners, Kit Galloway and Sherrie Rabinowitz, set up a televised link with a sister café, the Metropophobia Café in Phoenix, Arizona, so that the evening's lecture could span electronic space. Dave Blackburn of Virtual Ventures in Los Angeles had convened the group and invited me to give the first address. Brett Leonard, a frequent contributor to the café, had written and directed the movie *Lawnmower Man*. Because his movie celebrates the special effects of VR, I took the opportunity to link my philosophical writings to this place of popular culture.

■ ■ ■

One sharp observer—my son—once described my underlying approach as "Techno-Taoism." In short, my analysis accepts a certain degree of inevitability about our embrace of technology while I try to incorporate a deliberate balance, a balance of energies learned over years from Taoist practices. So, if pressed for a single thread running underneath these essays, I might admit to "Techno-Taoism," but my purpose is not to fashion a style of this or that but to illuminate certain phenomena, to go more deeply into where we are and where we are headed.

Not only do we have a breakthrough in computer interface, but even more important, we now face the challenge of knowing ourselves and determining how the technology should develop and ultimately affect the society in which it grows.

At the same time, I offer these essays in the hope that my beloved *philosophia* will awaken from her slumber and once again radiate brightly and move beautifully as she has in past centuries and in my dreams.

Long Beach, Calif.   M. H.
December 1992

ACKNOWLEDGMENTS

This book draws on the thoughts, projects, and ideas of many people. I cannot possibly mention everyone who has contributed to the work, but what follows comes from my recollection of the last few years.

I had the good fortune early on to make the acquaintance of several pioneers in the VR (virtual reality) field. Their work in telepresence, teleoperators, and virtual reality fed my speculations. Foremost among these is Myron Krueger, the "father of artificial reality," who states his ideas about art and science with clarity and force; both Dr. William Bricken and Meredith Bricken at the Human Interface Technology Lab of the Washington Technology Center share the passion of fellow philosophers; Dr. Chris Esposito at Boeing Aircraft outfitted me for my first virtual helicopter flight; both Michael Naimark and Robin Minard gave me day-long discussions of their VR projects at the Banff Centre for the Arts; and also at Banff is Dr. David Rothenberg, whose writings in philosophy and electronic music corroborated the direction of this book. Professor Michael Benedikt instigated the First Conference on Cyberspace at the University of Texas in May 1990, and he furthers constructive thinking at every chance he gets. Dr. Sandra Helsel of the Meckler publishing company summoned the First Virtual Reality Conference in San Francisco in December 1990, and she understands the role of philosophy in cultural life. Howard Rheingold taught me a lot from his lookout post on the WELL and from his books. Dr. Bob Jacobson tirelessly hosts the network group sci.virtual-worlds from the University of Washington and he is a boon to everyone in the VR community.

Dr. David Zeltzer of the MIT Media Lab gave me a crucial insight into the impulse behind VR; Dr. Frederick Brooks, Dr. Henry Fuchs, and Warren Robinett at the University of North Carolina shared many ideas and aspirations; Dr. Jonathan Waldern of W Industries was open and candid about his view of future VR developments; Dr. Joseph Henderson, M.D., director of the Interactive Media Laboratory at the Dartmouth Medical School, was kind and showed me the wide range of human issues involved; and Dr. Michael Moshell and Jacki Morie at the Institute for Simulation and Training at the University of Central Florida were most hospitable and encouraging.

I learned much from joint discussions with Sandy Stone and Brenda Laurel. I also bow to my friendly help in Japan from Koichi Murakami at Fujitsu Corporation, Ryoji Nakajima at Matsushita, Haruo Takemura at Advanced Technology Research, and Roe Adams III at RPG Cyberspace. Drs. Stephen Ellis and Lewis Hitchner, both at NASA–Ames, sat me down for a philosophical chat. My thanks also to Jaron Lanier and George Zachary of VPL, who see the need for philosophical reflection on their field. Colonel Richard Satava, M.D., of the U.S. Army Medical Corps, kindly corrected some of my misconceptions about telepresence. Dr. Mike Zyda, a computer scientist at the Naval Postgraduate School, provided help and information on several occasions. Tom Barrett of Electronic Data Systems has been a friend ever since we met at the First Conference on Cyberspace. Drs. Charles Grantham and Brad Smith from the University of San Francisco brought new breadth in applying VR concepts. Both Dr. Richard Economy of GE Aerospace and Dr. Pete Tinker of Rockwell Science Center clarified virtual environments for me. Dr. Emily Howard at Rockwell advised me at crucial moments, and Randal Walser—especially Randal Walser—of Autodesk Incorporated made it clear to me over the years that philosophers belong to the team of cyberspace creators.

Dr. John Latta, president of 4th Wave, revealed to me the business side of the endeavor, while MacCagie Rogers and his Mythseeker System showed me the sublime side; David

Smith, president of Virtus Corporation, was always a friend of this project (while still secretly harboring a preference for Tom Clancy's take on technology). Dr. Michael Century and Douglas MacLeod, directors at the Banff Centre for the Arts, lavished time and interest on the philosophy of VR. Dave Blackburn, president of Virtual Ventures, invited me to address the first meeting of the Southern California Virtual Reality Special Interest Group, and Kit Galloway and Sherrie Rabinowitz provided the electronic hookup. Florian Brody and Bob Stein, president of Voyager Company, demonstrated what could be done and dreamed in interactive multimedia.

I want to acknowledge the many dear and wonderful people who joined me since 1991 to form the Long Beach Virtual Reality Group, including Dr. Roman Yanda, Chris and Linde Farmer, Dr. Kent Palmer, Dr. Ralph Lewis, Merrie Martino, Sheila Finch, Margaret Elliott, Roger Trilling, Laurence Rozier, Dr. Phil Agre, and Dr. Dave Warner. The Chamber Pot Literary Society, led by Dr. Lenny Koff at UCLA, gave me feedback on several occasions. Bettijane Levine of the *Los Angeles Times* probed my work with frequent questions, as did Vera Graaf from the *Süddeutsche Zeitung*. Ben Delaney of *CyberEdge Journal* encouraged my writing, as did Phil Bevis of Arundel Press, and Karin August of VRASP extended my network. Dr. David Weinberger at Interleaf Systems supported my work with his writing skill and his intellectual background. Eugene Mallon in Totnes, England, and Greg Panos, in Lakewood, California, showed me new connections. At an important point, Jeanne Ferris brought her enthusiasm to the dawn of this book.

My friends in academia include Drs. William McGowan and Helmut Wautischer, who gave me the precious gift of hours of fellowship and understanding. Dr. Paul Tang showed continued interest in my projects, as did the Scheler expert, Professor Manfred Frings. Drs. Leslie Burkholder and Robert Cavalier at Carnegie-Mellon University expressed their interest on behalf of the American Philosophical Association Committee on Computers and Philosophy. Professor Don Jones at the University of Central Florida provided a forum in which I could address colleagues in academia. Dr. Richard Lanham of the UCLA Writing Program showed me the parallels in our work. Professor Henning Pryds and Thomas Moeller Kristensen at Odense University gave me an opportunity to test my ideas at the Danish Humanities Institute. My patient and perceptive editor at Oxford University Press, Cynthia A. Read, helped organize this book.

I also want to acknowledge Mark Mitchell and Tom Huchel at Technology Training Corporation and at the Education Foundation of the Data Processing Management Association. As their VR consultant, I learned much from organizing and chairing a number of conferences in Washington, D.C. The conferences gave me a continual update on the actual state of the art and brought home the importance of keeping the public discussion of technology on a serious and sober level.

Finally, thanks to all who tossed ideas to me through electronic mail on Internet at the address mheim@beach.csulb.edu or on CompuServe at ID: 76645,2075 or on the WELL at mheim@well.sf.ca.us. I continue to welcome ideas and comments from readers and netdwellers.

CONTENTS

environment, a virtual reality that we can enter. How far can we enter cyberspace and still remain human?

Contents

THE METAPHYSICS OF
VIRTUAL REALITY

■ 1

INFOMANIA

What is the state of the English language? No state at all. It
is in process. Our language is being word processed. If languages have states of health, sick or well, then ours is manic.

We face a tidal wave of written words. The wave of future shock swells on the horizon. First came speed reading,
a twentieth-century version of literacy. Next photocopy dupli-
cation, the word processor, and the fax machine. Now we
drive a technology that drives our verbal life faster and faster.
The word processor is computerizing our language.

Word-processed submissions have doubled the work load
of editors at commercial and academic presses. Writers grow
prolix, with manuscripts bloated to twice normal size. The
prose is profuse, garbled, torturously disorganized, as if the
difference between writing and revising were passé. Pages
are becoming more difficult to read. Reams of paper pour out
unedited streams of consciousness. The only writer who admitted that he was no faster than he had been before computers was Isaac Asimov, who published 141 books in 138
months.

Before 1980 the microcomputer was a crude, costly kit
for hobbyists and experimenters. Then Dan Bricklin and Dan
Fylstra created software for an electronic spreadsheet (an ac-
counting tool for figuring finances in rows and columns). *Vis-
icalc* ran on the Apple II and opened the market for desktop
computers. In 1981 International Business Machines (IBM) per-
suaded businesses that computerized spreadsheets would increase productivity. Once installed, computers could also

run other software, including word processing. The lure of greater productivity hooked professional writers, too, and by now most writers use word processing.

IBM coined the term *word processing* in 1964 to describe a brand of typewriter. The magnetic tape "Selectric" typewriter (MTST) boasted word-processing capabilities because it used magnetic tape to store pages of text. You could select pages for retrieval from electronic memory, which greatly streamlined the production of texts. Machines dedicated solely to word processing, like the Wang, soon appeared. The quantum leap in writing technology, however, came with microcomputers. The broad base of microusers allowed word-processing software to flourish. A decade earlier, data processors had used text-editing software on mainframe computers to create programs for number crunching. Their editing programs applied information-processing techniques rather than allow direct human interaction with texts on video monitors. When video arrived, inventors like Doug Engelbart and Ted Nelson saw that computers could do more than aid mechanical typewriting. They believed that word processing could amplify mental powers and increase our command over language. Word processing thus ceased being a typing gadget and became a cultural phenomenon. Over 80 percent of computer use is now word processing. Today computers spew out the bulk of written English.

During the 1980s a new vocabulary established the computerization of English. To be initiated, you had to repeat buzzwords like *access, input,* and *output.* You learned to speak of *files* having no apparent physical dimensions, *menus* offering a selection of nonedibles, and *monitors* providing vigilance over your own words. You learned to navigate with *wrap around* and a *cursor*—sometimes dubbed *cursee* as it became the recipient of your profanities. You may have even explored *mouse compatibility,* the *ASCII code,* and the difference between *RAM* and *ROM* memory. At the very least, you addressed yourself to *floppies* and *windows,* to *function keys* and program *documentation* (read: instruction manual). You had to take into account *block moves, hyphenation zones,* and *soft return* versus *hard return.* The editorial *cut-and-paste* became yours electronically. You learned not only to *delete* but also to *unerase*, then to *search-and-replace*, and onward to *globally search-and-replace. Automatic formatting* and *re-formatting* entered your writing routine.

Once initiated into the basics of word processing, you sigh, "This is bliss!" No more cutting paper and pasting, no more anxiety about revisions. Now you can get to work without the nuisance of typing and retyping. Words dance on the screen. Sentences slide smoothly into place, making way for one another, while paragraphs ripple rhythmically. Words become highlighted, vanish, and then reappear instantly at the push of a button. Digital writing is nearly frictionless. You formulate thoughts directly on screen. You don't have to consider whether you are writing the beginning, middle, or end of your text. You can snap any passage into any place with the push of a key. The flow of ideas flashes directly on screen. No need to ponder or sit on an idea—capture it on the fly!

But the honeymoon fades, and the dark side of computing descends upon you. The romance with computers shows its pathological aspects: mindless productivity and increased stress.

Your prose now reads—well—differently. You no longer formulate thoughts carefully before beginning to write. You think on screen. You edit more aggressively as you write, making changes without the penalty of retyping. Possible changes occur to you rapidly and frequently, so that a leaning tower of printouts stretches from the wastebasket to the heights of perfection—almost. The power at your fingertips tempts you to believe that faster is better, that ease means instant quality.

Business in America embraced computers under the magic rubric of *productivity.* Yet company reports do not seem to get better after thirty drafts. Real economic productivity in the United States actually declined over the last decade, and so has the competitiveness of the U.S. economy. Feel productive; push more paper.

Universities and colleges also bought into computerization. Miles of fiber-optic cable make subterranean links between academic buildings, snaking under the tree-lined footpaths like invisible superhighways. Yet few believe that computer networks actually advance liberal learning or that a greater outpouring of scholarly research makes better readers. Push a button; fell a tree.

Before computers, newspaper editors had a mentor–apprentice relationship with reporters. Reporters would write a piece and show it to their editor; after blue-penciling it, the editor would discuss it with the reporters. The reporters would then take responsibility for making the changes. Now things are different. The editor gets the electronic text, makes the changes, and then sends the reporters a copy. Reporters are not learning how to rewrite. While the editors are becoming better writers, the reporters are becoming data entry clerks.

If your company had a computer network installed, you could conduct business without worrying about coordinating schedules or reserving conference rooms or flying on airplanes to meetings. Through electronic mail, you can belong to a virtual (nonphysical) work group. You exchange reports or PROF notes without the small talk of phone conversation and the delay of paper mail. To prevent the accusation "You didn't tell me you were going to do that," you (and everyone else in the group) just hit a key and copies of your message fly off to everyone in the network.

On the receiving end, however, life is less rosy. Computer-generated notes, memoranda, and reports accumulate. Files clutter your work space, daunting your mind with their sheer volume. You are working in an intellectual swamp. Because you do not know immediately which files are worth saving, you have to wade through each of them before deleting any. The paperless office has more, not less, junk mail.

Physical hazards also lurk on the dark side of computing. Phosphorescent words on the screen hold a hypnotic attraction. So intensely do they attract that human eyes blink less often when viewing computer texts. The cornea of the eye requires frequent fluid baths, and eyelids normally bathe and massage the eyeballs by blinking every five seconds. But the stress of computer interaction tends to fix vision in a stare. As blinking decreases, the eye muscles have difficulty focusing.

The resulting strain eventually leads to refractive error, most often myopia.

The stress of digital writing causes more than myopia. Because it is intensely interactive and yet nearly frictionless, computer work involves more prolonged strain than does pencil or typewriter. You take fewer rest breaks. You have no file cabinets to visit, no corrections to make by hand, no variety of physical motions. Fingers just keep moving, repeating the same keystrokes. You hardly notice your unrelieved adaptation to the machine's specifications. The result is a workplace epidemic called *repetitive motion syndrome* (RMS). The inflamed hand and arm tendons of RMS patients often require surgery, and doctors are finding permanent damage to bodily movement in many RMS patients. The word processor is not merely a glorified typewriter.

I have yet to find a single writer who learned word processing and then abandoned it for pen or typewriter. Most writers and journalists share Stephen White's affection for word processing. In *The Written Word* he says of his fellow craftsmen:

■ A writer of any kind who does not work on a word processor is either dead broke or some kind of fool: it is as simple as that and we should not shilly-shally about it. He may be at the same time an absolutely first-rate writer, but although he may well dispute it, he gains nothing by his abnegation, and only makes life harder for himself and, to a limited extent, for others.[1]

Gore Vidal, neither broke nor a fool, would disagree. "The word processor is erasing literature," he says.[2]

But make no mistake. Despite the pathology of infomania, with its mindless productivity and its technostress, computers are here to stay. Efficiency, speed, and networked communication are in our bones. Our life rhythm moves to the tempo of the computer.

Already in 1957 Martin Heidegger noticed a shift in the felt sense of time. He saw the drive for technological mastery pushing into the human interior, where thought and reality meet in language:

■   The language machine regulates and adjusts in advance the
mode of our possible usage of language through mechanical
energies and functions. The language machine is—and above
all, is still becoming—one way in which modern technology
controls the mode and the world of language as such. Meanwhile, the impression is still maintained that man is the master
of the language machine. But the truth of the matter might well
be that the language machine takes language into its management, and thus masters the essence of the human being.[3]

Heidegger's philosophy was neither Luddite nor technophobic.
He resisted every attempt to categorize his view of technology
as either optimistic or pessimistic. Whether the glass was half-
empty or half-full, Heidegger was interested in the substance
of its contents. He was a soft determinist, accepting destiny
while studying different ways to absorb its impact.

Word processing is part of our destiny. Each epoch has
its love affair, its grand passion, an enthusiasm that gives it
distinction. Pyramids or cathedrals do not distinguish us, and
shopping malls will never last. Ours is not the age of faith
or reason but the age of information. Madness, Plato reminds
us, is ambivalent; it can be divine or insane, inspired or crack-
pot. Lovers, inventors, and artists are maniacs. So are computer enthusiasts. For infomaniacs, word processing is not
merely a tool.

Language technology belongs to us more essentially than
any tool. When a technology touches our language, it touches
us where we live. The chief inventors of word processing were
aware of this. These visionaries were not marketing a commercial product but seeking a revolution in the way we think.
They wanted to alter radically the way we interact with language. In the 1960s Douglas Engelbart wrote "The Augmentation of Man's Intellect by Machine" as he put together the
first text-processing hardware and software at the Augmentation Research Center (ARC). He balked at the inflexibility of
the means we have for handling symbols. If we could manipulate symbols in tandem with computers, he argued, we could
boost thought processes at least as much as handwriting
boosted the powers of preliterate humans. Engelbart was in

fact not trying to replace the mind with artificial intelligence. Instead, he conceived of the computer as a symbol manipulator for supercharging thought processes at the language level. Computers could constitute a world network in which the thoughts of countless individuals merge. Since Engelbart, many others have introduced software to affect our prose composition, our word choices, and even our logical processes.

Computer networks can be revolutionary. The 1989 pro-democracy uprisings in China were supported by computer networks and fax machines connecting thousands of Chinese people around the world. Computer bulletin boards created a public forum for free expression. Government-suppressed news streamed into China from outside. In Beijing, calls for freedom and reform circulated in Tiananmen Square on computer printouts.

Literature, too, changes as the written word migrates to electronic text. On computers, literature presents an unlimited cross-reference system for all symbol creations. A text includes footnotes that open up onto symphonies, films, or mathematical demonstrations. Browsing means push-button access to the text of all texts, or *hypertext*, as Ted Nelson called it. Hypertext and its offspring HyperCard are already evolving non-linear ways of reading. Books like Joyce's *Finnegans Wake* deserve another look as hypertext. Hypertext heightens non-linear and associative styles. Background knowledge and commentary pop up at the touch of a button. Like fractal structures, a text can turn back on itself linguistically, and hypertext shows the turns, the links, the recurring motifs, and the playful self-references.

How will traditional books fare? When Heidegger looked again ten years later in 1967, he saw a rising crest of information that, he suspected, might soon swallow his own writings: "Maybe history and tradition will fit smoothly into the information retrieval systems which will serve as a resource for the inevitable planning needs of a cybernetically organized mankind. The question is whether thinking, too, will end in the business of information processing."[4] He saw a growing obsession with data without a concern for significance.

Writing is the primary means we have for putting our thoughts before us, for opening mental contents to criticism and analysis. Using computers for writing, we experience language as electronic data, and the machines reinforce information over significance.

Information is a unit of knowledge that by itself has only a trace of significance. Information presupposes a significant context but does not deliver or guarantee one. Because context does not come built in, information can be handled and manipulated, stored and transmitted, at computer speeds. Word processing makes us information virtuosos, as the computer automatically transforms all we write into information code. But human we remain. For us, significant language always depends on the felt context of our own limited experience. We are biologically finite in what we can attend to meaningfully. When we pay attention to the significance of something, we cannot proceed at the computer's breakneck pace. We have to ponder, reflect, contemplate.

Infomania erodes our capacity for significance. With a mind-set fixed on information, our attention span shortens. We collect fragments. We become mentally poorer in overall meaning. We get into the habit of clinging to knowledge bits and lose our feel for the wisdom behind knowledge. In the information age, some people even believe that literacy or culture is a matter of having the right facts at our fingertips.

We expect access to everything NOW, instantly and simultaneously. We suffer from a logic of total management in which everything must be at our disposal. Eventually our madness will cost us. There is a law of diminishing returns: the more information accessed, the less significance is possible. We must not lose our appreciation for the expressive possibilities of our language in the service of thinking.

Notes

1. Stephen White, *The Written Word: And Associated Digressions Concerned with the Writer as Craftsman* (New York: Harper & Row, 1984), p. 68.

2. Gore Vidal, in *New York Review of Books*, March 1984, p. 20.
3. Martin Heidegger, "Hebel—Friend of the House," trans. Bruce Foltz and Michael Heim, in *Contemporary German Philosophy*, ed. Darrel E. Christensen (University Park: Pennsylvania State University Press, 1983), vol. 3, p. 95.
4. Martin Heidegger, Preface to *Wegmarken* (Frankfurt: Klostermann, 1967), p. ii. [Author's translation]

■ 2

LOGIC AND INTUITION

How does thinking at the computer differ from thinking with paper and pencil or thinking at the typewriter? The computer doesn't merely place another tool at your fingertips. It builds a whole new environment, an information environment in which the mind breathes a different atmosphere. The computing atmosphere belongs to an information-rich world—which soon becomes an information-polluted world.

First, the files you create grow rapidly, forming an electronic library of letters, papers, and other documents. Through on-line connections, you save pieces from the work of colleagues and friends, notes about future projects, and leftovers from database searches. Add some serendipitous items to disk storage—maybe the Gettysburg Address, the Constitution, or the King James Bible—and you find yourself soon outgrowing your disk-storage capacity. CD-Roms then spin out encyclopedias, the *Oxford English Dictionary,* or the entire corpus of ancient Greek literature. As the load of information stresses your mental capacity, you sense that you've come down with infomania.

Because the computer helped generate all this information, you naturally hope that the computer will in turn help mop up the mess. The computer can indeed hack a neat pathway through the dense information jungle. Computer data searches find references, phrases, or ideas in an instant, in the nanoseconds it takes the microprocessor to go through huge amounts of data. A word processor or database takes a key phrase and in a flash snaps a piece of information into view. So there you are, lifted by the computer out of the morass generated by computers. You can search through thousands of periodicals in minutes, without ever having to know anything about silicon microchips, high-level code, or sorting algorithms. All you need is some elementary search logic that you can learn in about an hour. Today most computer searches use elementary Boolean logic.

What is Boolean logic? Alfred Glossbrenner in *How to Look It Up Online* describes Boolean logic in terms simple enough for most computer users: "AND means a record must have both terms in it. OR means it can have either term. NOT means it cannot have the specified term." Glossbrenner chides those who belabor the complexities of Boolean logic and bewilder the user: "You sometimes get the impression that the authors would be drummed out of the manual-writers union if they didn't include complicated discussion of search logic laced with plenty of Venn diagrams—those intersecting, variously shaded circles you learned about in sophomore geometry. Forget it!"[1]

But alas, what Glossbrenner wants us to forget will soon enough slip into oblivion as technology enfolds us in its web of assumptions. Frequent reading and writing on computers will soon allow us little distance from the tools that trap our language. They will fit like skin. The conditions under which we work will grow indiscernible, invisible to all but the keenest eye. Present everywhere like eyeglasses on the end of our noses, computers will hide the distortion they introduce, the vivid colors they overshadow, the hidden vistas they occlude. Like microscopes, computers extend our vision vastly, but unlike microscopes, computers process our entire symbolic life, reflecting the contents of the human psyche. Boolean search logic and other computer strategies will soon enough become second nature for literate people, something they take for granted.

What people take for granted was once something startling and unprecedented. A felt transition like the present alerts us to the change, and so we have an opportunity to ponder the initial shifts in the life of the psyche. We can ask, How does Boolean search logic affect our thought processes and mental life? What dark side of infomania is hiding behind those "intersecting, variously shaded circles you learned about in sophomore geometry"?

The significance of Boolean search logic deserves far more than a sidebar in how-to manuals. Boolean logic, displayed graphically by the circles of the Venn diagrams, constitutes a central achievement of modern logic. Modern logic, which makes the computer possible, got its footing in the work of Gottfried Leibniz (1646–1716), whose discoveries laid the foundations of computer systems and the information age. So when we inspect Boolean logic for its side effects, we are looking at the implicit heart of the world we inhabit. Boolean logic functions as a metaphor for the computer age, since it shows how we typically interrogate the world of information.

Humans have always interrogated the world in a variety of ways, and each way reveals a distinct approach to life: Socrates pushed for personal definitions; Descartes and Galileo taught scientists to pose questions with empirical hypotheses; McLuhan teased our awareness with his enigmatic slogans; Heidegger drew on a scholarly history of reality; and Wittgenstein worried over odd locutions. The type of question we ask, philosophers agree, shapes the possible answers we get. The way in which we search limits what we find in our searching.

Today we interrogate the world through the computer interface, where many of our questions begin with Boolean terms. The Boolean search then guides the subconscious processes by which we characteristically model the world. Once we notice how computers structure our mental environment, we can reflect on the subconscious agencies that affect our mental life, and we are then in a position to grasp both the potential and the peril. So let's return again to those simple Venn diagrams from sophomore geometry and to the Boolean logic on which they are based.

George Boole (1815–1864) discovered the branch of mathematics known as *symbolic logic*. Boole's "algebra of logic" uses formulas to symbolize logical relations. The formulas in algebraic symbols can describe the general relationships among groups of things that have certain properties. Given a question about how one group relates to another, Boole

Logic and Intuition

could manipulate the equations and quickly produce an answer. First, his algebra classifies things, and then the algebraic symbols express any relationship among the things that have been classified—as if we were shuffling things in the nested drawers of a Chinese puzzle box.

Take two referential terms, such as *brown* and *cows*: all objects that are brown = B; all objects that are cows = C. An algebraic formula can represent the relationship between these two terms as a product of mutual inclusion: "All brown cows" = BC. For more complex formulas, add a logical NOT ($-C$) as well as an AND (BC and $C - B$). Once you know that (BC and $C - B$) = F (where F means any animal that "lives on the farm"), you can conclude that $BC = F$ or also that any cow, no matter what color, lives on the farm. You can build up terms that represent a whole series of increasingly complex relationships, and then you can pose and calculate any implication from that series. You can even make symbolic formulas represent a very long chain of deductive reasoning so that the logical form of each part of the argument rises to the surface for review and criticism, making it possible to scan an argument as if it were a mathematical problem.

Historically, Boole's logic was the first system for calculating class membership, for rapidly determining whether or not something falls into one or another category or class of things. Before Boole, logic was a study of statements about things referred to directly and intuitively at hand. After Boole, logic became a system of pure symbols. Pre-Boolean logic focused on the way that direct statements or assertions connect and hold together. A set of statements that hangs together can be a valid deductive pattern. Validity is the way that conclusions connect with their supporting reasons or premises. The traditional study of logic harked back to Aristotle, who first noticed patterns in the way we assemble statements into arguments. Aristotle called the assemblage of statements *syllogisms*, from the Greek for a pattern of reasoning. Aristotle himself used symbols sparingly in his logic, and when he did use symbols, they served merely to point out language patterns. Aristotle's symbols organized what was already given in direct statements. With Boolean logic, on the contrary, direct statements have value only as instances of the relationships among abstract symbols. Direct language becomes only one possible instance of algebraic mathematics, one possible application of mathematical logic.

Boole inverted the traditional relationship between direct and symbolic languages. He conceived of language as a system of symbols and believed that his symbols could absorb all logically correct language. By inverting statement and symbol, Boole's mathematical logic could swallow traditional logic and capture direct statements in a web of symbolic patterns. Logical argument became a branch of calculation.

The term *symbolic logic* first appeared in 1881 in a book by that title. The book's author, John Venn, introduced the first graphic display of Boole's formulas. Venn continued Boole's plan to absorb the direct statements of language into a general system of abstract algebra. With mathematics as a basis, Venn could solve certain logical difficulties that had perplexed traditional Aristotelian logicians. With mathematical precision, modern logic could present linguistic arguments and logical relationships within a total system, a formal organization having its own axioms and theorems. Systemic consistency became more important than the direct reference to things addressed in our experience.

Note already one telltale sign of infomania: the priority of system. When system precedes relevance, the way becomes clear for the primacy of information. For it to become manipulable and transmissible as information, knowledge must first be reduced to homogenized units. With the influx of homogenized bits of information, the sense of overall significance dwindles. This subtle emptying of meaning appears in the Venn diagrams that graphically display Boolean logic.

The visual display that John Venn created begins with empty circles. Venn noted how Boolean logic treats terms, like *brown* and *cows*, strictly as algebraic variables and not as universal terms referring to actually existing things. In Boole's logic, terms function like compartments or drawers that may or may not contain any actual members. Boole's logic can use terms that are empty, the class of unicorns, for example. A term with no actually existing members is a null set, an empty compartment. As modern logicians say, the terms of logic do not in themselves carry existential import. The terms reveal relationships among themselves, but they remain unconnected to existence or to the direct references of firsthand experience. (Mathematics also shares this existential vacuum: $2 + 2 = 4$ remains mathematically true whether or not four things actually exist anywhere.) Boolean logic uses terms only to show relationships—of inclusion or exclusion—among the terms. It shows whether or not one drawer fits into another and ignores the question of whether there is anything in the drawers. The Boolean vocabulary uses abstract counters, tokens devoid of all but systemic meaning.

On Venn diagrams, then, we begin with empty circles to map statements that contain universal terms. We can map the statement "All the cows are brown" by drawing two overlapping circles: one representing cows and the other, brown things. Shade in (exclude) the area that represents cows and that does not overlap the area representing "brown things," and you have a graphic map of the statement "All the cows are brown." The map remains accurate regardless of whether or not any cows actually exist; you could equally well have drawn a map of the unicorns that are white. Add a third circle to represent spotted things, and you can map "No brown cows are spotted" or "All brown cows are spotted," and so on.

What does this procedure really map? According to Boolean logic, no cows or brown things or spotted things need actually exist. All we have mapped is the relationship between sets or classes. The sets could refer to custards or quarks or square circles.

In its intrinsic remoteness from direct human experience, Boolean search logic shows another part of the infomania syndrome: a gain in power at the price of our direct involvement with things. The Boolean search affects our relationship to language and thought by placing us at a new remove from subject matter, by directing us away from the texture of what we are exploring.

To add particular statements to our map, like "Some spotted cows are brown," we need to introduce more symbols. We can map statements about particular things on the dia-

grams by stipulating another conventional symbol, often a
star, an asterisk, or some other mark. Statements that imply
that a particular member of a class actually exists must be
specifically marked as such; otherwise, the general term labels
a potentially empty compartment. From the outset, then, Boolean logic assumes that as a rule, we stand at a remove from
direct statements about particular things in which we existing
beings are actually, personally involved.

This shift in the meaning of logical terms has drastic
consequences for logic itself and for logic as a formal study.
Traditional Aristotelian logic presupposed an actual subject,
ideal or real, to which logical terms or words refer. Traditional
logic also presupposed that logical thinking is, like spontaneous thought and speech, intimately involved with a real
subject matter. Mathematical logic gained the upper hand by
severing its significance from the conditions under which we
make direct statements. Today, logicians like Willard Van
Orman Quine can argue that a concrete and unique individual
thing (to which we refer as such) has no more reality than
"to be the value of a variable," at least when we consider
things "from a logical point of view." The modern logical
point of view begins with the system, not with concrete content. It operates in a domain of pure formality and abstract
detachment. The modern logical point of view proceeds from
an intricate net of abstract relations having no inherent connection to the things we directly perceive and experience.

We can contrast this aloof abstraction with the traditional
logic that still swam in the element of direct experience. Traditional logic began with direct statements, insofar as its logical language presupposed as necessary the existential
interpretation of statements. When we state something in everyday language, we attribute something to something; we
attribute the color mauve to the wall, the quality of mercy
to a creditor. We speak of what is before us, and we speak
in the context of other people who may also have access to
what we are talking about. We commonly assume the existence or at least the existential relevance of what we are talking about. Modern symbolic logic, on the contrary, mimics
modern mathematics, which has no interest in the actually

existing world, not even the world of direct statements. In this sense, modern logic operates at a remove from our everyday involvement with things.

But why pick on modern Boolean logic? Don't all logics bring abstraction and alienation? Even the words we use to pose any question testify to a gap between us and the wordless subject we are thinking or talking about. Any logic can distance us. We sometimes run across a person arguing with impeccable logic for a conclusion contrary to our own gut feelings, and we often feel overwhelmed, and forcibly so, by the sheer power of the argument itself. Logic can move like a juggernaut adrift from any personal engagement with its subject matter. Someone with a great deal less experience, for example, can make us feel compelled to accept a conclusion we know instinctively to be wrong. We feel the logical coercion even though we may have much more familiarity with the matter under discussion. Arguing with someone like Socrates or William F. Buckley can be disconcerting. We sense a line of thought pushing inexorably through the topic, perhaps even in spite of the topic. Logic, like mathematics, operates outside the intuitive wisdom of experience and common sense. Hence the mathematical idiot savant. Like math, logic can hover above particular facts and circumstances, linking chains of statements trailing from some phantom first premise. We can be perfectly logical yet float completely adrift from reality. By its very nature, logic operates with abstractions. But modern logic operates with a greater degree of abstraction than does Aristotelian logic, placing us at a further remove from experience and from felt insight.

When college students study those Venn diagrams from "sophomore geometry," they feel the pain of that disengaged logic. When they first learn to symbolize statements and arguments in symbolic logic, they must pass through a lengthy and painful process of converting their English language into abstract symbols. So far removed does this logic stand from the direct everyday use of language that the textbook refers to the process of converting arguments into symbols as "translation." Before analyzing their thoughts logically, students must translate them to fit the system of modern logic. Statements in direct English must first undergo a sea change.

The painful translation into symbols signals acute infomania. But when logic works on the computer, this pain turns into convenience. When the computer automatically and invisibly converts input into algebraic bytes, the user is shielded from the translation into modern logic. Instead of the human mind puzzling over how language fits the system, the computer does the fitting; it transforms our alphabet into manipulable digits.

As a medium, the computer relieves us of the exertion needed to pour our thoughts into an algebraic mold. The shift from intuitive content to bit-size information proceeds invisibly and smoothly. On the machine level, the computer's microswitches in the central processing unit organize everything through a circuit based on symbolic logic, and Boolean searches simply apply that same logic to text processing. Hardly noticing this spiderlike, nondirect logic, we stand at a new remove from concretely embedded language. The computer absorbs our language so we can squirt symbols at lightning speeds or scan the whole range of human thought with Boolean searches. Because the computer, not the student, does the translating, the shift takes place subtly. The computer system slides us from a direct awareness of things to the detached world of logical distance. By encoding language as data, the computer already modifies the language we use into mathematized ASCII (American Standard Code for Information Interchange). We can then operate with the certitude of Boolean formulas. The logical distance we gain offers all the allure of control and power without the pain of having to translate back and forth from our everyday approach to the things we experience.

But so what if computer power removes us from direct intuitive language? So what if Boolean logic injects greater existential distance from practical contexts than any previous logic? Don't our other text tools also operate at a remove from direct context-embedded language? Isn't any medium, by definition, a mediation? If the Boolean search operates at a great remove from direct oral discourse, don't also pen and paper, not to mention rubber erasers and Linotype typesetting machines?

Nonlinguistic tools, like erasers, do indeed insert a distance between ourselves and our context-embedded mother tongue. And, yes, using a rubber eraser does affect us in subtle, psychological ways. Teachers understand that getting a student to use an eraser marks a significant step on the road to good writing. A self-critical attitude distinguishes good from bad writing, and picking up an eraser means that we are beginning to evaluate our own words and thoughts.

But using Boolean search logic on a computer marks a giant step in the human species's relationship to thought and language. Just as the invention of the wax tablet made a giant stride in writing habits, forever marginalizing chiseled stones, so too Boolean search logic marks the new psychic framework of electronic text woven around us by computers. With electronic text we speed along a superhighway in the world of information, and Boolean search logic shifts our mental life into a high gear.

The Boolean search shows the characteristic way that we put questions to the world of information. When we pose a question to the Boolean world, we use keywords, buzzwords, and thought bits to scan the vast store of knowledge. Keeping an abstract, cybernetic distance from the sources of knowledge, we set up tiny funnels to capture the onrush of data. The funnels sift out the "hits" triggered by our keywords. Through minute logical apertures, we observe the world much like a robot rapidly surveying the surface of things. We cover an enormous amount of material in an incredibly short time, but what we see comes through narrow thought channels.

Because they operate with potentially empty circles, the Boolean search terms propel us at breakneck pace through the knowledge tunnel. The computer supports our rapid survey of knowledge in the mode of scanning, and through the computer's tools we adapt to this mode of knowing. The scanning mode infiltrates all our other modes of knowing. The byte, the breezy bit, and the verbal/visual hit take the place of heavier substance.

Of course, the computerized reader doesn't pluck search terms out of pure air. The funnels we fashion often result from a carefully honed search strategy. In *How to Look It Up Online*, Glossbrenner advises the reader:

■ Meditate. Seriously. You may not be a Ninja warrior preparing for battle, but it's not a bad analogy. If you ride in like a cowboy with six-guns blazing, firing off search terms as they come into your head, you'll stir up a lot of dust, expend a lot of ammunition, and be presented with a hefty bill but very little relevant information when you're done. . . . *Think* about the topic beforehand. Let your mind run free and flow into the subject. What do you know and what can you extrapolate about the subject?[2]

What Glossbrenner calls meditation actually works to serve calculation. What he describes is no more than a deep breath before taking the plunge. Meditation of this kind only sharpens an already determined will to find something definite. The user meditates in order to construct a narrower and more efficient thought tunnel. But even if we build our tunnels carefully, we still remain essentially tunnel dwellers.

The word *meditate* came originally from the Latin *meditari*, meaning "to be in the midst of, to hover in between." The meditation that Glossbrenner prescribes—prudent advice as far as it goes—helps the user zero in more closely on a target. It is the fill-up before a drive on the freeway, not the notion to hike in the countryside.

If we in fact take inspiration from the ninja warrior, we should recall Kitarō Nishida's teachings about "the logic of nothingness" (*mu no ronri*). The ninja warrior empties his mind before battle precisely by abandoning all specifics, by relaxing his attention so that the windows of awareness open to fresh perceptions. Genuine meditation refreshes our original potential to move in any direction. Our highest mind remains alert but flexible, firm but formless—in short, omnidirectional. Meditation truly expands the psyche and opens it to the delicate whisperings of intuition.

A Taoist sage once wrote that "thinking is merely one way of musing." Tightly controlled thought remains but a

Logic and Intuition

trickle in the daily stream of thoughts flowing through the psyche. Most of the time, the background mind muses with a soft undercurrent that quietly sorts things out, gently putting things together and taking them apart. We do our best thinking when sitting before the fireplace on a crisp winter night or lying on the grass on a balmy spring day. That's when our minds are most fully engaged, when we are musing.

Computer-guided questions sharpen thinking at the interface, but sharpness is not all. A more relaxed and natural state of mind, according to Siu, a Taoist, increases mental openness and allows things to emerge unplanned and unexpected. Rather than sharpen the determined will, we must preserve a state of no-mind in which our attention moves free of the constricted aims of consciousness. The musing mind operates on a plane more sensitive and more complex than that of consciously controlled thought. Musing is not wild in the sense of wanton but wild in the sense of flowing, unforced, and unboundedly fruitful. Thinking itself happens only when we suspend the inner musings of the mind long enough to favor a momentary precision, and even then thinking belongs to musing as a subset of our creative mind.

Now contrast the Boolean scan with a meditative perusal through traditional books. The book browser moves through symbols in the mode of musing. Books do in fact have a linear structure that unfolds sequentially, page by page, chapter by chapter, but seldom do readers stick to reading in this way. When we look something up in books, we often find ourselves browsing in ways that stir fresh discoveries, often turning up something more important than the discovery we had originally hoped to make. Some of our best reading is browsing. The book browser welcomes surprise, serendipity, new terrain, fresh connections where the angle of thought suddenly shifts. The browser meditates every moment while under way, musing along a gentle, wandering path through haphazard stacks of material. The browser forgoes immediate aims in order to ride gently above conscious purposes, in order to merge with an unexpected content in the pages. The browser feels wilderness beckon from afar.

The Boolean reader, on the contrary, knows in advance

where the exits are, the on-ramps, and the well-marked rest stops. Processing texts through the Boolean search enhances the power of conscious, rational control. Such rationality is not the contemplative, meditative meander along a line of thinking, that the search through books can be. The pathway of thought, not to mention the logic of thought, disappears under a Boolean arrangement of freeways.

The Boolean search treats texts as data. When you search a database, you browse through recent material, often covering no more than the last ten years. Cutting off the past in this way streamlines the search. But a musing cut off from historical roots loses the fertile exposure to false starts, abandoned pathways, and unheard-of avenues. An exclusive focus on the recent past curtails our mental musings, and a narrow awareness sacrifices the intuitive mind.

Boolean search logic affects our mental vision just as long hours at the computer screen affect our eyesight. In a relaxed state, our eyes accept the world passively as a spectacle of discovery. Only when we strain to see do our eyes lose the surprise of perceptions. Constant straining induces a sensory myopia in which we need to strain in order to see better what we wish to see. We lose much of our peripheral vision when we use our eyes willfully. Likewise with the mind's eye. A relaxed and easy thought enjoys intuitive turns, and thinking at its best muses over human symbols. Boolean search logic cuts off the peripheral vision of the mind's eye. The computer interface can act like the artificial lens that helps us persist in our preconceptions. Boolean logic can unconsciously entrench us in our straining ways, hurting us as much mentally as the carpal tunnel syndrome hurts us physically. We may see more and see it more sharply, but the clarity will not hold the rich depth of natural vision. The world of thought we see will be flattened by an abstract remoteness, and the mind's eye, through its straining, will see a thin, flattened world with less light and brightness.

But notice how we do in fact always use some holistic guesswork, even when we are trying out best to shut off the mind's peripheral vision. Our Boolean searches could never begin without vague hunches and half-seen surmises. We need hunches and inklings to start with. Unfortunately, the Boolean search places our hunches in the service of a skeletal logic far removed from the direct operations of language.

If computers aid our searching minds, we must not abandon the books during our leisure time. The serendipitous search through books is necessary for knowledge and learning. Browsing often evokes daydreams and unsuspected connections; analogies and pertinent finds happen among the stacks of physically accessible pages. Although not as efficient as the Boolean search, library browsing enriches us in unpredictable ways. Looking for something in a book library frequently leads to discoveries that overturn the questions we originally came to ask.

Book libraries hold unsystematic, unfiltered collections of human voices and thoughts. Libraries are repositories not so much of information as of the intuitions of countless authors. The books in libraries remain physical reminders of the individual voices of the authors, who often speak to us in ways that shock and disturb, in ways that break through our assumptions and preconceptions, in ways that calm and deepen. The word *museum* derives from the Greek word for the Muses, goddesses of dream, spontaneous creativity, and genial leisure. Libraries may be, in this strict sense, the last museums of the stored language, the last outposts of predigital intuition.

Today libraries are becoming information centers rather than places for musing. The Los Angeles County Public Library, the world's largest circulating library, receives more requests for information than requests for books. In 1989, one university in California opened the first library without books, a building for searching electronic texts. Books still remain a primary source, but they are rapidly becoming mere sources of information. A large volume of book sales doesn't necessarily prove that the book, with its special psychic framework, endures as such. Many books today gain attention as nonbooks linked to cinema, television, or audio recordings.

Searching through books was always more romance than busyness, more rumination than information. Information is by nature timebound. Supported by technological systems, information depends on revision and updating. When books become mere sources of information, they lose the atmosphere of contemplative leisure and timeless enjoyment. Old books then seem irrelevant, as they no longer pertain to current needs. One of the new breed of information publishers epitomizes this attitude in a pithy warning: "Any book more than two years old is of questionable value. Books more than four or five years old are a menace. OUT OF DATE = DANGEROUS."[3]

As book libraries turn into museums of alphabetic life, we should reclaim their original meaning. Museums are places for play, for playing with the muses that attract us, for dreams, intuitions, and enthusiasms. Information plugs us into the world of computerized productivity, but the open space of books balances our computer logic with the graces of intuition.

Notes

1. Alfred Gloss Brenner, *How to Look It Up Online* (New York: St. Martin's Press, 1987), p. 109.
2. Ibid., p. 116.
3. Daniel Remer and Stephen Elias, *Legal Care for Your Software* (Berkeley, Calif.: Nolo Press, 1987), p. ii.

▰ 3

HYPERTEXT HEAVEN

What is hypertext? *AskSam* might help. *AskSam* is a textual database, a program for manipulating written materials. The program also provides hypertext on IBM-compatible computers. The *AskSam* reference manual defines nearly every aspect of the program. When it comes to hypertext, however, the manual mumbles: "A facility." On-line help screens within the program are more forthcoming: hypertext is "a facility by which all text on screen becomes a point-and-shoot menu for commands." Hardly crystal clear. Translated, the help screen means that you can place the cursor on any word or phrase and call up all the other contexts containing that same word or phrase.

AskSam does speedy searches, a three-hundred-page book in twenty seconds. A good word processor can nearly match that speed. The hypertext search is different because it is nodal, or relational. Any phrase or group of phrases in a textual database can be matched or compared with everything else in the database. The user need not stop with a single search. Put the cursor over two words, and a screen will show every context in which those words appear, even if the contexts occur at opposite ends of the text. Move the cursor to another interesting phrase in these contexts, and more screens will come up, displaying other contexts with same phrase, and so on. Each word or phrase is a key to other references to itself. At any time you can view a backtrack list or history that traces all the references you followed. Hypertext is a mode of interacting with texts, not a specific tool for a single purpose. You can realize what hypertext is—or can be—only by sitting down with it for half an hour. Once caught in the interactive nature of the thing, you can begin to imagine an immense range of possible applications.

In 1987, Apple Computer brought out the first hypertext commercially available on computers. HyperCard on the Macintosh holds files ("stacks") that resemble electronic index cards. Unlike index cards, however, the stacks are relational, or automatically cross-referenced with one another. Because the stacks are electronically linked, they allow instant cross-referencing. Stacks link everything in a text or in a whole group of texts. Texts then become a hypertext in which everything in the text relates to everything else in the text. In other words, hypertext is a dynamic referencing system in which all texts are interrelated. Hypertext is no less than electronic intertextuality, the text of all texts, a supertext.

The term *hypertext* refers to the existence of an unnoticed or additional dimension. In board games and in mathematical physics, the term *hyper* means "another dimension." Hyperchess refers to chess played on a board with more than two dimensions. Not only are the chess pieces three dimensional, but the rules of play recognize the existence of the third dimension. In mathematical physics, *hyperspace* means "space with more than three dimensions." If the three-dimensional Euclidean space of the universe is curved back on itself, it becomes a limited but unending hypersphere. When written words and phrases have an extra dimension, they are like crystals with infinite facets. You can turn over an expression and view it from any number of angles, each angle being another twist of the same text. Words and phrases appear juxtaposed or superimposed. The sense of a sequential literature of distinct, physically separate texts gives way to a continuous textuality. Instead of a linear, page-by-page, line-by-line, book-by-book approach, the user connects information in an intuitive, associative manner. Hypertext fosters a literacy that is prompted by jumps of intuition and association.

The intuitive jump in hypertext is like the movement of space ships in futuristic fiction. When this fictional travel exceeds the speed of light, it becomes a jump through hyperspace. At such speeds it is impossible to trace the discrete points of the distance traveled. In one of his science fiction novels, *The Naked Sun*, Isaac Asimov depicts movement in hyperspace like this:

■ There was a queer momentary sensation of being turned inside out. It lasted an instant and Baley knew it was a jump, that oddly incomprehensible, almost mystical, momentary transition through hyperspace that transferred a ship and all it contained from one point in space to another, light years away. Another lapse of time and another Jump, still another lapse, still another Jump.[1]

Like the fictional hyperspace, hypertext unsettles the logical tracking of the mind. In both, our linear perception loses track of the series of discernible movements. A hypertext connects things at the speed of a flash of intuition. The interaction with hypertext resembles movement beyond the speed of light. Just as computer outlining weakens the fixed hierarchy of traditional script and print outlines, so hypertext supports the intuitive leap over the traditional step-by-step logical chain. The jump, not the step, is the characteristic movement in hypertext.

When first reading hypertexts, certain works of literature come immediately to mind, in particular James Joyce's *Finnegans Wake*, whose style is a foretaste of hypertext. *Finnegans Wake* spins nets of allusions touching myriad other books and often alludes to other parts of itself. Its complex self-references and allusions have daunted and frustrated many a reader: few books outside the Bible call for so much background knowledge and so much outside commentary. The secondary literature on the *Finnegans Wake* is enormous, with glossaries of puns and neologisms and etymologies of the many foreign and concocted words. More important, this book embodies the structural shape of hypertext. It is the ne plus ultra of nonlinear and associational style, a mess of hidden links and a tangle of recurring motifs. Joyce worked on *Finnegans Wake* for over seventeen years, in a nonlinear fashion not unlike the way a person typically uses a word processor. The book was not created with a beginning, then a middle, and finally a conclusion. Rather, Joyce produced sections as the muses moved him. Sometimes he wrote only a single large word across the page in crayon (he was nearly blind at the time). Yet everything in *Finnegans Wake* dovetails like a woven pattern, turning back on itself linguistically like a wave of fractal structures. When Gerrit Schroeder and Tim Murphy began computerizing Joyce's grand linguistic dream at UCLA in 1987, they realized that the hermeneutic structure of the novel matches hypertext. The two were meant for each other. A grand and puzzling work of the twentieth century seems to break out of its book format to find a second life on computers. As the century passes away, *Finnegans Wake* presages a reincarnation of human symbols.

For some students of literature, hypertext offers a practical test for the theories about literature. One theory of literature, for example, proposes the view that every written text is a spin-off of some other already-presented aspects of cultural life. No book invents the language or idiom it uses. The strongest advocates of this theory of intertextuality even argue that all the images that a reader receives in reading a book derive exclusively from other preverbal cultural activities. A text therefore is an aggregate of other texts, especially if we allow texts to include the audio and video arts. Hypertext gives us a way of experimenting with this theory, perhaps even helping prove or disprove it.

One professor turned his classroom into a laboratory to test the thesis. Using *Guide* hypertext on Macintosh computers, he had his students work with texts by Dryden, Pepys, Milton, and Spratt, all written in 1667. The students used their computers to determine whether each of the texts could be connected, by word or by concept, with any of the others. The conditions of the experiment limited the certainty of the results, but some students were apparently able to find notions in the texts that could not be traced back to any other texts. These notions were perhaps original and free of intertextual influences. One student, for instance, found references to open marriage in Dryden's play *Secret Love*. No other works of the period contain such a notion, even when checked against the books in the library. Hypertext thus offers ways of testing theories of language and looking at them anew.

With its jumps and its linked allusions, hypertext first emerged as an idea in the 1960s. The term appeared in the speculations of Ted Nelson. Struggling to write a philosophy book, the young Nelson ran into trouble when trying to organize his writing in the form of a conventional book. In an interview in April 1988, Nelson recalls:

■ I've always been a generalist. I got into computers because I had trouble writing my books on philosophy. I had a complete philosophical system at the age of twenty-three. When I say complete, I mean it was comprehensive in the sense of being well-articulated and tied together. It was still lacking, but I knew I was on to something, and I had terrible trouble organizing it. But I knew that organizing ideas was a hard problem. Many people confronted with big manuscript situations say, "it must be me," or "the material isn't there." Neither of these was the case. So the problem was a hard problem, and that's what got me into computers. I took a computer course and said Wow! This is the way to organize reading and writing in the future![2]

I define hypertext as nonsequential writing with free user movement. It has nothing to do with computers logically; it has to do with computers pragmatically, just the way large numbers and large bookkeeping schemes have nothing to do with computers logically but, rather, pragmatically.

Nelson originally planned his philosophy as a linear system, something fit for a book. He was struggling to squeeze a complete system into the book format. One assumption he naturally made was that a philosophy is, or should be, systematic.

The assumption that philosophy should be systematic is a recent one. It became widespread only in the last three hundred years or so, which—as basic assumptions go—is the day before yesterday. The systematic format won favor with the rise of rationalism in the seventeenth and eighteenth centuries. Spinoza wrote his ethics as a system of geometrical proof. Kant laid out his critiques like blueprints for architecture, and Fichte and Schelling followed suit. Hegel fit all essential truths into a small encyclopedia. Each of these philosophers tried to present a total thought system in which he spelled out the whole truth so explicitly that the main principles of the system rule over every single part. They tried to articulate things totally.

Philosophy was not always bound to systems in this sense. Before the seventeenth century, thinkers did not strive for that kind of total system. Consider one philosopher-theologian who often comes to mind as an example of a premodern systematic thinker. Thomas Aquinas wrote summaries of his teachings, the *Summa Theologica*. On closer inspection, however, his approach to a system through summaries is not very systematic in the modern sense. The system in Aquinas's writings is subordinate to other concerns. In his *Summa*, he poses and answers questions; he cites and interprets ancient texts; he recounts at length opposing views and argues with them; and he raises objections to his own views. In short, Aquinas creates a dialogue with his culture, not a closed system. In this way, the book format of the *Summa* is incidental. The bound volumes are collections. They channel a continuing discussion, and they reproduce the style of an oral dissertation or a spoken defense. So the dominant format for philosophical thinking was not always the book. With the arrival of popular printed books, the demand for the single-minded system arose. Philosophy as a closed system was fit for books.

When Nelson connected his philosophy book with computers, he did not realize that in making this connection he was undermining the book he thought he was trying to write. He made the modern assumption that philosophy should proceed systematically. At first Nelson imagined a text sprinkled with dynamic footnotes, or what he called *links*. He conceived of these links as no more than footnotes organized by computer, as simply electronic references. The footnotes would enfold subordinate parts of the system as well as the references to other books. At first Nelson did not think of the footnotes as jumps. Rather he conceived of a literature linked by footnotes that would make any number of secondary texts available with electronic speed. A book would then have links with the other texts to which it explicitly referred.

When activated, the electronic link brings reference material immediately to the screen. Computers do the footwork. The reference can be a paragraph, an article, or an entire book. It can even be, Nelson later realized, a film or a photograph or an audio recording. A return key brings the user back to the point in the original text where the link flag appears.

The hypertext link turns out in fact to be much more than a reference tool. The link indicates the implicit presence of other texts and the ability to reach them instantly. It implies the jump. With the jump, all texts are virtually coresident. The whole notion of a primary and a secondary text, of originals and their references, collapses. In magnetic code there are no originals, no primary, independently existing documents. All texts are virtually present and available for immediate access. The original text is merely the text accessed at the moment, the current center of focus. Computerized links eventually pushed Nelson to speak of hypertext as "electronic publishing in an ever-growing interconnected whole."[3] The notion of isolated verbal information collapses, too. The stored writing is once again wed to images, as had been the case with illuminated manuscripts and oriental calligraphy. This time, the stored images are animated and attached to recorded sound.

As the computers' memory banks grow and as the communication satellites carry the data, hypertext is growing toward a comprehensive network. Two decades passed before the software and the hardware could support Nelson's vision even to a limited extent. For twenty years, the available hardware could put no flesh on the notion of hypertext. It remained a fantastic vision. But Nelson held on to it, insisting all along that computer software and hardware remain merely pragmatic considerations. The crucial thing, he maintained, is the new design for literature. Text design outweighs the imperatives of hardware and software. Computers serve a strictly pragmatic role insofar as they help create new designs for reading and writing. The important thing is the conscious creation of a format for texts. Besides calculators and intelligence simulators, computers are workstations for more flexible ways of designing texts. Computer hardware is important only because it helps us realize a more flexible literacy. Using an older terminology, you might say that computers are efficient causes in fulfilling a final cause: the mutation of text.

Yet Nelson had not gone deeply enough into the hidden core of computers. The historical origin of the computer shows that it was more than an instrument to serve an arbitrary purpose. The original computer already contained the spark of hypertext as an idea. Only now does the truth of this idea dawn on us. And this truth is not a recent invention. Underneath the computer's calculating power lies an inner core sprung from a seed planted two centuries ago. By staying with the primacy of text design, Nelson unknowingly preserved the core, the inner *telos*, the half-forgotten origin of computers. That initial germ for the birth of computers started with the rationalist philosophers of the seventeenth century who were passionate in their efforts to design a world language.

The notion of a world language arose in the early modern period. Gottfried Leibniz (1646–1716) founded modern logic as the science of symbols. With his rationalism, Leibniz placed his stamp on the modern mind-set. His work presaged hypertext, the total text, intertextuality, the text of all texts. But Leibniz's work contained the concept of hypertext only seminally. Although he constructed computer prototypes, Leibniz was able to build machines suitable only for numeric calculation. Not even at this time could Leibniz's machines do deductive proofs with his binary computational logic. Only centuries later—after Boole, Venn, Russell, Whitehead, Shannon, and others—could symbolic logic handle both deductive proofs and electronic circuits. Later, John von Neumann could use Leibnizean binary numbers to develop digital computers. But what Leibniz lacked in hardware, he made up for in speculative imagination.

Leibniz's speculations revolved around language. On the practical side, he was a courtier, a diplomat, and an ecumenical theologian. He strove to unify the European world. His idealism brought him to believe that by focusing on the new physical sciences, the national states of Europe could unite under a shared project. In an age of religious wars and growing nation-states, Leibniz imagined a world federation based on common linguistic symbols. He advocated a universal system of symbols for all the sciences, hoping that a rational scientific language might smooth the way toward international cooperation.

Leibniz believed all problems to be, in principle, soluble. The first step is to create a universal medium in which to communicate. With a universal language, you can translate all human notions into the same basic set of symbols. A universal character set (*characteristica universalis*) can absorb every significant statement or piece of reasoning into a logical calculus, a system for proving things true or false—or at least for showing them to be consistent or inconsistent. Through a shared language, many discordant ways of thinking can exist under a single roof. Once disagreements in attitude or belief are translated into matching symbols, they can yield to logical operations. Problems that earlier seemed insoluble can stand on a common ground. In this belief, Leibniz was to some extent continuing a premodern Scholastic tradition. That medieval tradition believed that human thinking (in its pure or ideal form) was more or less identical with logical reasoning and argument. To the partisans of dispute Leibniz would say, "Let us put this into our common language, let us sit down and figure it out, let us calculate." He worked on a single system to encompass all the combinations and permutations of human thought. He longed for symbols to foster unified scientific research throughout the civilized world. The universal calculus would compile all human cultures, bringing human languages into a single shared database.

Standing behind Leibniz's ideal language is a premodern model of human intelligence, a model that measures humans against a being who knows things perfectly. Human knowledge models itself on the way that a divine or an infinite being knows things. Finite beings go slowly, one step at a time, seeing only moment by moment what is happening. On the path of life, a finite being cannot see clearly the things that remain behind on the path or the things that are going to happen after the next step. A divine mind, on the contrary, oversees the whole path. God sees all the trails below, inspecting at a single glance every step traveled, what has happened, and even what will happen on all possible paths below. God views things from the perspective of the mountaintop of eternity. Human knowledge, thought Leibniz, should emulate this *visio dei*, this omniscient intuitive cognition of the deity.

No temporal unfolding, no linear steps, no delays, limit God's knowledge of things. The temporal simultaneity, the all-at-once-ness of God's knowledge serves as a model for human knowledge in the modern world as projected by Leibniz's work.

The power of Leibniz's modern logic made traditional logic seem insignificant by comparison. Aristotle's logic was taught in the schools for centuries. Logic traditionally evaluated the steps of thought, valid or invalid, as they occur in arguments in natural language. Modern logic absorbed the steps of Aristotle's logic into a system of symbols, and it became a network of symbols that could be applied to electronic switching circuits as well as to arguments in natural language. Just as non-Euclidean geometry can make up axioms that defy the domain of real circles (physical figures), so too modern logic freed itself of any naturally given syntax. The universal logic calculus could govern computer circuits. What impact this modern logic has had on everyday language and thought is still an unanswered question that needs more study.

One result seems certain. As the code that connects computer programs to the logic of circuits, modern logic smoothes the way for the postmodern jump of thought.

Hypertext emulates a divine access to things. Although God does not need to jump, the hypertext user leaps through the network of knowledge in something like an eternal present. The user feels intellectual distances melt away. Empowered by hypertext, however, the human victory over time and space is a merely symbolic victory. Human users remain on the level of symbols, as they are not really gods and do not see things in a simultaneous present all at once. Total information is the illusion of knowledge, and hypertext favors this illusion by letting the user hop around at the speed of thought.

Just as Leibniz believed he could absorb and not undermine logical thinking with his universal language, and just as Nelson believed he could preserve and not destroy his philosophy book with hypertext, so also the users float on the illusion that this hypertext style of reading surpasses without undoing all the earlier styles of reading. Hypertext users feel their minds reveling in intuitive, associational thinking. The more the routine sequences and chores belong to the computer program, the more the human psyche can give rein to immediate insights and creative angles. This trend is in keeping with our general cultural tendency to prefer fast images to the more intricate patterns of internal thought.

If the jump gains dominance over logical steps, hypertext literacy may epitomize the postmodern mentality. Contrasted with the older literacy, the hypertext jump may well belong to the postmodern condition described by the critic Charles Newman:

> Post-Modern means the first culture in history totally under the control of 20th century technology, and the first in five hundred years in which information is codified in ways which do not depend on literacy. All we have discovered, thanks largely to mass communications, is that "reality" is often more hollow than is verbal configuration, and that meanings date even faster than style.[4]

The information age may seem increasingly nonliterate or hollow because we move in quite a different symbolic element that has its own tempo. In contrast with literate cultures of the past, we face an enormous volume of stored information. Our ancestors faced the task of slowly learning from experience, of gleaning from life whatever they could discover. They then tried to amplify whatever they could confirm. They learned how to pass along an expanding knowledge base. They stored knowledge impersonally in writing and eventually learned to automate knowledge so that it could become information.

Today our task seems to be the reverse of that of our ancestors. Given a constant stream of information, we must be skeptical of any structures that narrow its flow. For us, no single overarching order can set up proper channels for the incoming information. We also need to sort through and make sense of the tide of information. Information is abundant but without any fixed center around which to organize it. Our task is to hold onto the anchor of our own experience to find meaning in the sea of information.

Hypertext helps us navigate the tide of information. In skipping through hypertexts, we undergo a felt acceleration of time. If computers cause impatience with finite human experience, then the term *hyper* in hypertext starts to remind us of another one of its cognate meanings. In psychology, medicine, and the social sciences, the prefix *hyper* means "agitated" or "pathological." Hypertext thinking may indeed reveal something about us that is agitated, panicky, or even pathological. As the mind jumps, the psyche gets jumpy and hyper. We can only hope that the postmodern hyperflood will not erode the gravity of experience behind the symbols, the patient, painstaking ear and eye for meaning.

Notes

1. Isaac Asimov, *The Naked Sun* (New York: Ballantine, 1957), p. 16.
2. Theodore Holm Nelson, interview with David Gans, *Microtimes*, April 1988, p. 36.
3. Ibid.
4. Charles Newman, *The Post-Modern Aura: The Act of Fiction in an Age of Inflation* (Evanston, Ill.: Northwestern University Press, 1985), p. 187.

■ 4

THOUGHT PROCESSING

The printing press and the computer are two very different ways of presenting knowledge. One works mechanically; the other, electronically. Today they belong together. The personal computer in the last decades of the twentieth century builds on Johann Gutenberg's printing press of 1457. Both broadly affect the way we preserve and convey knowledge.

How does the impact of computers differ from that of the printing press? For an answer we need first to disentangle the two, sorting out the print from the computer culture. Pulling them apart is not easy, especially so soon after the introduction of computers.

Print literacy was already in place when computers arrived on the scene. The personal computer entered our culture swiftly and with more fanfare than the printing press. Printed texts gave a firm footing to computers. Widespread reading material prepared the advent of electronic communications. Through the printed word, commercial culture promoted electronics rapidly and aggressively.

In its time, the printing press altered the format of written words. The first publishers needed to standardize punctuation, invent typefaces, and reduce books to sizes and styles appropriate to the mass market. (Mass market hardly describes the readers of the first printed books, who bought Greek, Latin, and Italian classics.) While the content printed by the first publishers was not revolutionary, the increased access to knowledge was. No longer rare, books became affordable. Under the economics of printing, literate people could possess a personal or family library. The fifteenth-century scholar Erasmus marveled at the Venetian publisher Aldus Manutius: "Aldus is creating a library which has no boundaries other than the world itself."[1] Printed books made the world a library, with a bookshelf in every home.

Four hundred years later, the world libraries bulged with printed materials. By 1950 there was talk of a knowledge explosion. Piles of printed material towered beyond reach, and mountains of paper grew inaccessible, ever more difficult to comprehend. Computers promise delivery from the Babel of books. Magnetic storage on computers gives faster access to texts and reduces physical storage space. Electronic text offers high-speed search and nearly instantaneous transmission. Once installed, the computer assigns to print an ancillary role, making it input or output for computers. Liberated from mechanical print, knowledge becomes a software configuration. Stored for faster access, printed words become information. The increased access speed means a change in quality. The computer has epistemological implications. It affects the way we represent what we know. Knowledge comes across differently, so that absorbing and thinking about information change. For one thing, the computer inserts a piece of knowledge into a network of information that includes everything else that is known. Knowledge then revolves in an informational loop.

One example of an informational loop occurs on the level of word processing. Writing is our primary means for putting our thoughts before us, for opening mental contents to critical analysis. In earlier ages, writing amounted to carving characters into rock or forming letters on parchment. The writing materials resisted inscription. You almost had to think through an entire sequence of ideas before committing it to writing. Writing traditionally meant composing ideas in your head, the habit of mentally formulating an ideational sequence. Composition traditionally meant composing your mind or putting together your ideas. The mind learns to hold ideas in mental sequence; which fosters linear thought processes: we try to start at the beginning, create a middle, and then end at the end. Ordered logic becomes a norm for the thought process;

it even becomes synonymous with careful thinking. The trained mind experiences the thinking as a reasoning to pre-suppositions, premises, middle terms, and conclusions. With resistant materials, this mental feat is accomplished before the writing process begins.

Word processing makes thoughts flow more directly. Not only the mind but also the eye follows insights as they occur. Thoughts appear on the screen nearly as fast as they come to mind. Sequence can be imposed afterward. Later we not only see what we have mentally enunciated, but also feel free to rearrange, reorganize, and change what we thought or said we thought. Serious penalties like retyping or cutting and past-ing no longer plague the computer user. If good thinking means self-editing and self-criticism, then the computer speeds up the thought process while preserving the discipline of objectifying ideas. The mind's eye and the physical eye work together. The eye is wired to the brain via the computer, making a feedback loop between the mind and the written word. Word processing is not just a quantitative improvement in getting a job done more efficiently.

Although we still associate computers with printing and with transmitting knowledge on paper, we are now interfacing with our own thoughts. Software transforms knowledge be-yond the limits of printed writing. Word processing leads to the more fundamental activity of thought processing. To see what the computer is actually doing beneath routine work, we have to expand our vision. We have to look backward first, then forward. We stretch our vision forward by considering the latest enhancements of writing and thinking on computers. First we look backward to the modern origins of knowledge processing.

Peter Ramus was an obscure logician who lived from 1515 to 1572. As Walter Ong has shown, Ramus provides a key to understanding the way that print shaped modern cul-ture. The cultural movement begun by Ramus changed the course of education, first in Europe and then in the American colonies. The Ramist movement introduced a novel form of education that advanced rational clarity in the modern Carte-sian sense and sought to reorganize the age-old traditions of

Western logic and rhetoric. The motor of Ramism was the printing press, which empowered the Ramist movement by making it feasible to reproduce and distribute outlines of knowledge on paper. The educational reform of Peter Ramus signaled, as Ong says in the preface to *Ramus, Method, and the Decay of Dialogue*, "a reorganization of the whole of knowledge and indeed of the whole human lifeworld."[2]

Ramus advocated knowledge outlines. The printing press could reproduce any number of pages displaying graphic trees that present summaries of a body of knowledge. Each page is a skeletal outline of a subject arranged systematically, with the branches on the tree showing how the parts of the subject matter connect. The printed page thus becomes a chart of topics divided into dichotomies with their parts and interconnections made clearly visible. That is, the printed Ramist text is a visual encyclopedia of cultural literacy in which topics and their parts appear in a nutshell.

Ramus backed his work with venerable authorities. He promoted his outlines with neoclassical appeals to the philosophy of Plato and Aristotle. The intellectual order of the outlines harked back, however crudely, to the Platonic–Aristotelian ideal of *diaresis*, an analytical procedure for summing up and dividing a subject matter. (Vestiges of the procedure linger in the Linnean botanical classification of plants by genus and species.) The Ramist outlines did not accurately reflect classical philosophy but reduced intellectual content to something easily memorized. Ramus was a dedicated pedagogue. His knowledge outlines did not make ontological statements about the nature of reality. Instead, they reduced knowledge to facile facsimiles to aid memorization. Early modern outlines were a tool for streamlining education.

The Ramist educational reform supported the general thrust of modern thought. Walter Ong correctly characterized modern philosophy—exemplified in the work of René Descartes—as binary, visualist, and monological. Binary means that the modern thinker accepts as self-evident the need to divide and subdivide every subject into oppositions or dichotomies. (The binary division also occurred in ancient philosophy, in the Pythagoreans and in Plato. But Plato dwelt on the

dynamics of opposites, and the Pythagoreans held 10 to be
the perfect number.) Modern thought is visualist in insisting
that truths display spatial fixity and detail. Descartes carried
out the Platonic metaphor of the "mind's eye" by taking literally the visual aspect of truth. He admitted into the inner heart
of truth the modern technology of optical instruments: the
telescope, the microscope, and scientific gauges. Cartesian
philosophy was also monological, nondialogical. Descartes
couched his philosophy in soliloquies and meditations about
his personal struggle to throw off the disputatious methods
of the traditional schools. Truth in the modern Cartesian sense
appears not in discussions or disputes but in experiments.
Experiments must follow established rules or methods, and
no dialogue establishes the rules or methods. Rather, a formal
logic supposedly dictates the rules.

The Ramist reform advanced the binary, visualist, and
monological mind-set. Ramist outlines presented knowledge
as topics branching visually across a silent page. Individuals
could absorb truths printed in books at a remove from the
direct challenge of discussion. The printing press could reproduce hundreds of typeset outlines at mechanized speed, making the (simplified) principles of Cartesian philosophy
available to many. Students had a tool for memorizing material rapidly and conveniently. Learning now seemed easier.

The epistemological impact of printing was noticed only
gradually. Europeans were hardly aware of the shift in the
quality of knowledge, as the new knowledge format only subtly undermined the older oral world of discourse and rhetoric.
Whereas truth was once understood to be a matter for public
dispute, it now became the property of individual minds.
Whereas understanding was previously connected with audi-
ble verbal utterance, it was now conceived in graphic, spatial
terms. And whereas subtle differences once provoked argument, the effort was now to keep distinctions clearly in memory. The analytic graphic tree simplified knowledge, and the
printed outline gave modern thought an engine to transform
culture.

The modern fondness for outlines continues to the present. Teachers drill outlining in grammar school. Lawyers,

scientists, and technical writers use logical outlines and flow-charts. Business and industry convey complex ideas and hier-archies by means of outlines and charts. Scholars may disagree about the details of outlining a Platonic dialogue, but most agree on the importance of outlining. All assume—sometimes anachronistically—that outlines present the inner logical struc-ture of thought.

Today professional writers see writing as a search for lucid outlines even before writing a single word. Before creat-ing a document, the writer produces a document of another kind, an outline. The writer gropes for the best possible struc-ture, assuming that logical order carries the burden of mean-ing. Stephen White speaks for most writers when he says in *The Written Word*:

■ I honestly believe that no writer, faced with a task that will require more than a few minutes of actual writing, sets out to write without an outline. There are times when the outline is rudimentary and held wholly in the mind. . . . The outlining procedure is a process of creating order and coherence. The search for order and coherence is what we call *thinking*. When it employs the written word beyond the manufacture of an outline, it is called *writing*.[3]

Writers commonly identify thinking with outlining: outlining is basic thinking, whereas the actual writing is a distinct craft. This is the modern notion of thinking in which a visual, hier-archic order guides the mind.

What happens when computers supersede the printing press, when the outline is computerized? If—as Ong shows for the early modern period—the printed outline expresses and reinforces the mind-set of a whole life world, what does a metamorphosis of the outline imply for us? What might we learn about ourselves, about life in the information age? If Ramist educational tools signaled a shift in epistemology, what changes might we expect from the thought patterns we find on computers?

The outline has its origin in the modern drive to organize things visually in logical schemes. In recent years outlining has migrated to computers, where it gains renewed power and momentum. But computer outlines differ intrinsically from printed outlines. Computer outlines are not, strictly speaking, outlines. They are outliners. The term suggests the active or, better, interactive nature of outlines on computers. Another name for outlining is idea processing or thought processing.

The outlines of the Ramist movement gave us a retrospective index of what is modern. The modern period, most agree, has ended, and outlines have changed accordingly. Computer outliners register the postmodern; they give us prospective glance into the future.

On the computer, a specific program shapes the outline, and the program automates the procedures of outlining. It is, of course, the writer who actively uses both computer and program to create outlines. The process is machine interactive in ways unknown in modern printed texts. Outliners incorporate dynamic interaction: free-form text searches, on-line references, and hypertext facilities. All these belong to thought processing.

Thought processing brings about an active interface between humans and machine. The outliner differs radically from the printed outlines so central to the modern Ramist tradition. No longer passive in paper books, the computer outline forges an interactive give-and-take between the user and the computer. Outliners supercharge the fixed, spatial modern outline. The spatial anatomy of an ideational sequence now turns fluid, dynamic, unconstrained, and users sense the difference immediately. People familiar with grammar-school outlines lose their bias against outlining, their reluctance giving way to a feeling that computer outliners support their personal thought, fitting it like a second skin.

Each outlining program shows a different side of thought processing. *KAMAS* (Knowledge and Mind Amplification System) was the first outliner on personal computers. A group of former philosophy professors wrote the program, seeing outlining as a vehicle for fostering a certain way of thinking. Compusophic Systems wanted to promote the educational philosophy of the Chicago school of the 1940s and 1950s, which advocated the neoclassicism of Leo Strauss, Richard

McKeon, and Mortimer Adler. The group described KAMAS as the "classic vehicle" for thinking, as it helps writers structure their thoughts in hierarchical levels. The classification scheme of KAMAS has a remote ancestor in the logic of Aristotle. Aristotle's logic drew on a hylomorphic metaphysics that identifies reality with the generic nature of things and their species differences. KAMAS orders ideas as they come, by subordinating the more specific under the more general idea. The outliner ranks ideas by their generality. For instance, a set like

    Computer programs
    Operating systems
    Database
    Outliners
    ThinkTank

becomes

    Computer programs
        Operating systems
        Applications programs
            Text manipulation
                Word processors
                Outliners
                    ThinkTank
        Database

All in good Aristotelian style, KAMAS fosters logical hierarchy, but it does more than the genus–species approach suggests. What I cannot show here is the dynamic flexibility of electronic outlines, their interactive quality that invites reordering. Because it is electronic, KAMAS works neither classically nor neoclassically, but in the postmodern mode. Traditional outlines on paper are immovable. You do not interact with them. If you want to make additions or changes in hierarchy, you have to redo the entire outline or a good part of it. Electronic outliners are infinitely more flexible, even when structured for hierarchy.

Unlike KAMAS, most outliners openly advertise a post-modern thought process; they encourage an order that floats and shifts. Both *ThinkTank* (Living Videotext, Mountain View, California) and *PC-Outline* (Softworks Development, Mountain View, California) help plan and structure information. They function as textual databases in which you keep names, addresses, and projects. They are called *personal information managers* (PIMs). You enter information in the familiar structure of topics and subtopics, but you can hide lower levels of information until you want them revealed. You can see an outline in overview with only the top headings showing, or you can examine all the lower details by revealing subheadings and their accompanying text. You can instantly hoist a subtopic to topic level, or you can demote topics—all without having to redo the entire order.

Nothing remains rigidly fixed. Order itself is at your disposal. You can change the numbering of topics from the traditional roman numerals to "A . . . 1 . . . a . . . 1" or to the legal style "1 . . . 1 . . . 1 . . . 1 . . . 1" or to indentations with no numbers or to any graphic symbols. You can move paragraphs to any position, and the numbers change automatically. You can sort entries in alphabetical or ASCII order, ascending or descending. With PC-Outline, you can mark a word or number with ":" and then instruct the program to sort on the second, third, or later field. For instance, you can set off dates as sortable fields, as in the following:

1. KCRW Radio   :87-03-05   :(213) 450-5183
   Phone author, Joe Frank

2. Rightway Learning Centers   :86-12-05   :(213) 375-2005
   Talk to Director, Larry Manuel, Ph.D.

3. Alan Cranston, U.S. Senate   :87-05-12   :(213) 215-2186
   Urge a meeting with the Heal-the-Bay people

You can sort this group of headings by name, date, or by telephone number. Each sort is nearly instantaneous. The outliner becomes a miniature database, which is the fundamental way in which symbols exist in the information age. Written words become a textbase.

Outliners also do cloning. The same topic appears in more than one place, so that a change to a topic in one location makes changes in all the others. With PC-Outline, you can have up to nine outlines open on the screen at once, each in its own window, and you can move information among them. You can open and close each outline separately, and you can move and resize the windows separately. Multiple windows permit different kinds of work in the same work space. Outlining becomes a creative environment in which thinking, writing, and planning coexist in an outlined structure. Unlike modern outlines, outliners do not merely schematize a passively received volume of knowledge. Rather, the postmodern outliner is a working environment in which you strive constantly to think in order to produce.

Another program uses dynamic outlining to promote a software work environment. *Framework* (Ashton-Tate, Torrance, California) integrates several aspects of computerized writing: word processing, outlining, database management, electronic spreadsheet, and telecommunications. It organizes information by using an abstract element, the frame, to identify and put boundaries around each work fragment. Each frame is itself an outline, with every sublevel potentially another outline—like a Chinese box. You can manipulate each fragment within a frame: move it around, edit, add something, remove a part, combine it with other fragments, copy it completely or in part, print it, file it, or send it out the phone lines. You can develop ideas through words, graphs, spreadsheets, or databases, all within frames that you can stretch, combine, and arrange as building blocks for an infinite variety of structures. The outliner allows an overview of the work in progress as a whole, or it shows the detailed parts. Framework presents a complex architecture of information.

Outliners transform the unit for what constitutes an idea. The unit of ideas grows larger than the modern topic or the traditional notion in the mind, and the computerized idea is inseparable from graphics and numerical data. In an integrated software environment, text is juxtaposed with numbers, graphics, and telephone input. The heading of an outlined idea is the visible tip of a large packet of information. The postmodern version of the idea differs in quality because of the sheer quantity of information available.

Most outliners do not integrate as many functions as Framework does. Usually outliners specialize in a single aspect of thought processing. *Freestyle* (Summa Technologies, Kentfield, California), for example, is a real-time outliner. As you enter text, the word processor prompts for general headings. The time in which you compose the text is the same time in which the computer generates an outline. The outline is not the skeleton of a future document but a constantly abstracting anatomy of a living thought sequence. It shows the background steps of what you are writing. You can consult the outline at any point, either to examine a logical progression or to reorder it. The user assigns a hierarchy to topic and subtopic, but without prior organization. The program automates the process of abstraction.

The Freestyle outliner highlights another side of thought processing. The ultimate outliner would be a fully automatic program for text production. Outliners do in fact already exist that generate a partial text from a skeleton of ideas, with little human intervention. The complexity of a piece of writing obviously determines the necessary amount of human intervention. But research continues and at least suggests differences between traditional and electronic outlining. One structures knowledge for ordered recall, and the other simulates the dynamism of a thought process.

Some outliners still follow traditional logical structure by suggesting top–down thinking. They exhibit a major topic with subheadings and sub-subheadings. But other programs expose the novel logical structure beneath computer outlining. Rather than order implicative or subordinative relationships among ideas, the outliner turns the user's attention elsewhere, toward nodes of linked thoughts. The link or nodal jump replaces the logical step as the most characteristic thought movement.

Outliners like *Houdini* focus on complex relationships that are linked rather than subordinated. Houdini (Reliance

Company, Piedmont, California) allows you to connect any item to many inputs and many outputs. It fosters many-to-many connections. Instead of listing topics, Houdini connects one node to another by having its name placed on the other's input or output list. You can make the list of all nodes appear and check off those you want connected. Or Houdini can automatically find which links are common to a number of nodes and which nodes have the same inputs and outputs. A node is like a member of a genealogical tree. Parents are like input, and children are like output. Each member has two parents, each of those has two others, and so on. In a traditional outline, you would have to assign each descendant to one of their parents. In the usual computer outliner, you would have to use a cloning device. With Houdini, you can quickly list parents (as input) and children (as output). Here is a practical application of a simple link system:

One person used Houdini to classify her skills for a career change. She wanted to find work that would express as many of her skills and values as possible. She knew her skills and her values, and she needed to gather them into usable groups. Houdini's ability to connect different nodes allowed her to note the connections that were obvious. But because Houdini is flexible and quick, she was able to try out new patterns of connection much faster than she could have done with an outline processor or paper and pencil. The dissatisfaction she felt with her first clusterings led to experiments with new groupings of skills and values. In the end, she came up with a two-pronged career plan. An outliner's structure, with its one heading for many subheadings, would have discouraged the unusual combination she created.

The interactive process departs from hierarchical structure. Logical connections appear, and links are made only to the extent that a present-day offspring resembles one of its ancestors.

Other kinds of links derive from developments in modern logic, specifically Boolean logic. Boolean logic allows searches where words or phrases appear in combinations. Logical operators like AND and OR are used to combine the key search terms. The computer speedily produces all the contexts where that combination of keys is found. Programs like *GoFer* (Microlytics, East Rochester, New York) and *Text Collector* (O'Neill Software, San Francisco) search rapidly and often produce unexpected but useful insights into a piece of writing or collection of texts. Programs also offer searches with a so-called fuzzy logic that retrieves associative or cognate passages beyond the exact input of the user.

The logical link and fuzzy logic belong to another relatively recent breed of software: information managers or textual databases. In current information-managing software, links or nodal jumps replace logical steps. Rather than order the implicative or subordinative relationship between statements, this software allows the user's attention to glide freely over nodes of linked thoughts.

Information managers like *Information-XL* (Valor Software, San Jose, California) have been called "spreadsheets of the mind." This software is effective for keeping track of already-stored information. Information-XL has a link facility, that provides a permanent way of connecting information. Like the mind, this software cross-references items under multiple headings. Unlike the physical world, the mind does not limit the ways in which information can be arranged. Thinking about tomorrow, for instance, might spark a reminder of a meeting. A thought of the meeting might remind you of project X or of a person involved in the project. Linking allows information about a person to be connected with a date or a project as well as with a meeting. Information-XL does not copy the data, but instead the data are interpreted to be in two places at once. You can refer to the same information in as many different places as you need it. You cross-reference or link computerized text: changes in one instance of the link make changes in all instances of the data.

Information managers like Information-XL are close to becoming hypertext. As we have seen, hypertext differs from traditional printing as much as a movie differs from a canvas oil painting. Software not only accelerates our thought process, but also facilitates the birth of a new reality in which we think. We should not mistake the new digital reality for a neutral territory untouched by human intention. Software hides within it specific notions about how we do and how we should think within a digital environment.

Notes

1. "Aldus bibliothecam molitur, cuius non alia septa sint, quam ipsius orbis" (Erasmus, "Festina Lente," *Adagia*, in *Opera Omnia*, ed. Jean Le Clerc (1706), vol. 2, p. 103A.
2. Walter J. Onj, *Ramus, Method, and the Decay of Dialogue* (Cambridge, Mass.: Harvard University Press, 1958, 1983), p. viii.
3. Stephen White, *The Written Word: And Associated Digressions Concerned with the Writer as Craftsman* (New York: Harper & Row, 1984), pp. 118–19.

HEIDEGGER AND McLUHAN: THE COMPUTER AS COMPONENT

Heidegger and Computers

An odd juxtaposition? No philosopher highlights the clash
between technology and human values so sharply as Heidegger. Not only did he make technology central to metaphysics,
but he came to see in it the root evil of the twentieth century,
including the Nazi German catastrophe, which he described
as "the confrontation of European humanity with global technology." In both his life and his writings, Heidegger felt technology to be an overwhelming force that challenges the
reassuring maxims of traditional morality. Yet his death in
1976 did not permit him to see the century's most powerful
technological revolution: the proliferation of the microcomputer. He saw only the first glimmerings of computerization, the
mainframe dinosaurs of the computer age. But because his
work spans the gap between the days before computers and
the increasingly computerized present, Heidegger can become
a springboard for understanding the new situation of the sciences and the humanities.

The images we have of Heidegger the thinker, both pho-
tographic and mythic, place him in another time, another generation. In posed photographs, we see him sitting in a hut
on the quiet mountaintop of the Todtnauberg, surrounded

by shelves of books as he bends intently over a wooden writing table. The sun pours in the window. Under his pen, the manuscripts bristle with marginalia and scrawled notations of every kind, his pages a palimpsest heaped with layers of minute revisions. Heidegger the thinker is Heidegger the scholar, and the scholar searches ancient texts for clues to the history of Being. He looks for hints about where our essence, our heart, is today and whither the pull of the future.

This image of Heidegger feeds on nostalgia. Even the Heidegger of the photos, seated in his hut a half-century ago, working with pen on paper, had a keen sense of just how faded this picture was soon to become, how quickly this image would turn antiquarian. Because he connected being with time, Heidegger knew that reality changes and with it the task of thinking. He sensed the pace of change in the twentieth century, and he seemed to foresee what librarians realize today: "The image of the humanist scholar in the book-crammed study, thinking deep thoughts, will continue to be less and less viable in professional scholarship."[1] This recent observation by the director of a great college library confirms what Heidegger in his writings surmised: our rapid technological advance challenges the legacy of human thinking. Who better than the contemporary librarian knows the inner trend of today's scholar? Bid adieu to the "hochgewölbtes, engen gotischen Studierzimmer" of Goethe's Faust. The Schreibstube is giving way to the computer workstation, and scholarship requires a cybersage.

Computerized libraries already exist today without paper books, and by the year 2000, nearly every text of human knowledge will exist in electronic form. Heidegger sensed, with anguish, that his works would one day come to light in a world of scholarship that had grown alien to the meditative pathways that nurtured his thoughts. In 1967, he saw a rising crest of information that, he suspected, might soon engulf his own writings: "Maybe history and tradition will fit smoothly into the information retrieval systems that will serve as resource for the inevitable planning needs of a cybernetically organized mankind. The question is whether thinking, too, will end in the business of information processing."[2]

In his essay "The Age of the World Picture," Heidegger unearthed seeds planted by seventeenth-century Cartesian philosophy that have blossomed today as science merges with computer science.[3] The computer began to appear indirectly in Heidegger's mid-century writings as he took up the theme of calculative versus meditative thinking, for the computer was to become the supreme calculator.

The first time I ran across the conjunction of Heidegger and computers was in 1977 when Joseph Kockelmans returned from giving seminars in Europe. While in Trier, he made the acquaintance of two graduate students, Rainer Bast and Heinrich Delfosse, who were at the time breaking new ground in Heidegger studies. Kockelmans showed me some work from these two students by handing me a stack of computer paper twenty centimeters thick. It was a series of computer printouts listing the textual discrepancies among the various German editions of *Being and Time*. Since the 1960s, the computer analysis of texts had been used occasionally by humanists, but mainly to detect stylistic differences in classical works like those of Homer and Shakespeare. There in my hands lay the first discomfiting conjunction of Heidegger and computers. That computer printout eventually became the book *Handbuch zum Textstudium von Martin Heideggers "Sein und Zeit."*[4]

Until then, Heidegger and computers had been for me an odd juxtaposition, an abstraction under the heading "the question of technology." What my hands held was not an abstract treatise but a concrete, oxymoronic fact. Heidegger speculated on an all-enframing *Gestell* (technological system), ominous and threatening, but an abstraction looming like a metaphysical sphinx, terrorizing thought with a puzzling lack of specificity. Now here was computer text concretely manifesting that abstraction. The stack of printouts highlighted both the inevitability of a technologically informed scholarship and the soundness of Heidegger's fears that his work would soon become an object of technological scrutiny. Heidegger was now on computer. The question of technology had become the question about how to go about studying Heidegger.

Just what were the specific dangers of computers? At that time, the main philosophical answer to this question was what I call the computer as opponent. In this approach, the computer appears as a rival intelligence that challenges the human being to a contest.

The Computer as Opponent

In 1972, Hubert Dreyfus called attention to the danger of computers. Applying phenomenological analysis, Dreyfus argued that we must delineate carefully what computers can and cannot do, lest we become unrealistic about computers and fall into a misunderstanding of the kind of beings we ourselves are. In *What Computers Can't Do*,[5] Dreyfus observed how mid-twentieth-century culture tended to interpret the human being as an information-processing system. Researchers spoke of the brain as a heuristically programmed digital computer. Because the brain is a physical thing, Dreyfus noted, and because we can metaphorically describe the brain as "processing information," we easily slip into the unexamined dogma that human thinking operates in formal patterns and that properly programmed computers might be able to replicate these patterns. If computers could replicate thought patterns, might we not then justifiably say that computers think or have an artificial intelligence (AI)? Research funds were flowing into AI when Dreyfus raised his doubts. Dreyfus argued then—and continued to argue in his book *Mind over Machine*[6]—that we are deluding ourselves if we believe that we can create machines to replicate human thought. Dreyfus sought to establish the limits of artificial intelligence, and he saw the computer as a metaphysical opponent.

Most philosophical reasoning about computers still moves within the narrow confines of artificial intelligence, the computer as opponent: Is it possible for computers to think? Can human mental and perceptual processes fit the formulas of digital programs? How far can computers advance in simulating or surpassing human reasoning? Such are the questions that held, and still hold, the attention of philosophers from Hubert Dreyfus to John Searle. This line of inquiry goes only a short distance in exploring the existential questions raised by the conjunction of Heidegger and computers.

The computer opponent line takes for its paradigm the chess match. More combative than the Turing test, the chessboard places the human in a duel with the computer, the winner claiming superior intelligence. The game paradigm ensures that the relationship remains antagonistic. The combative paradigm still holds sway over the popular imagination, the human-versus-the-machine contest, with a winner/loser outcome.

Dreyfus first connected Heidegger with computers by working within this model. Observing an unbounded enthusiasm for artificial-intelligence research, Dreyfus drew on Heidegger's critique of technology to set limits on the kind of research that defines the human mind as an information processor. Dreyfus challenged the very idea that a chess-playing program "of any significance" could be built, and in 1965 he published a paper equating "Alchemy and Artificial Intelligence." This ruffled the AI researchers, and they took Dreyfus up on his challenge. In 1967, MIT researchers confronted Dreyfus with a computer chess program named Mac-Hack. To the delight of the AI community, Dreyfus lost to the computer in a public match.[7] Later, in *What Computers Can't Do*, Dreyfus explained his philosophical point—that he was concerned not with generic predictions but with the underlying comparison that hastily identifies intelligence with formal patterns or algorithms.

Dreyfus sharpened Heidegger's technology critique by focusing on the formal patterns that computers use. Because software programs run by explicitly stated instructions, the computer works on a level of intelligibility that Heidegger characterized as derived and not primordial. Formal patterns process reality but filter it through the screen of lucidity. What fails to fit the patterns gets lost in the process, even if we try to reintroduce the unknown into our interpretations. The tendency to interpret reality as essentially lucid or representable goes back to Plato, according to Heidegger's early reading of Plato.[8] Dreyfus sees in the computer, according to artificial-intelligence researchers, the apotheosis of metaphysics. Plato postulated the Good as subsistent in itself. The Good, the *agathon*, energizes the forms of things, making them stable and self-consistent. So too artificial-intelligence research, at least in one of its early phases, postulates formal patterns as the be-all and end-all of intelligence. (Much recent AI research is turning away from the priority of formal algorithms and instead is looking to "fuzzy logic.")

Dreyfus applied Heidegger's critique of technology to computers, but he conceived the computer too narrowly as an artificial-intelligence device. He saw the computer only as opponent. Yet the opposition of computer and mind/brain remains, as Heidegger would say, ontic rather than ontological. The two terms *mind/brain* and *computer/program* refer to beings, to definite entities within the world. We can compare and research the nature of these entities. We can investigate the causes of their operations, sizing up their powers and limitations, but still we treat them as beings, as entities delimited by their respective natures. The mind-versus-computer question is neither ontological nor existential. Whether or not the computer could, in principle, outsmart the mind or simulate consciousness—however intriguing a question this might be—does not touch what is happening to us through computerization. The chess paradigm distracts us from the present issue, because it makes us construe our relationship to computers as confrontational rather than collaborative.

Very different from the computer as an opponent is the computer as a component. The computer has become an ingredient in human knowing. Instead of confronting a potential rival, we find ourselves interfacing with computers. Computers are woven into the fabric of everyday life, and they have become an important thread in the texture of Western civilization. Our daily reliance on computers affects the way our culture proceeds, in everything from architecture to zoology. Instead of regarding computers as opponents, we collaborate with them. Increasingly rare is a computer-free stance from which to regard the computer as a separate device. Even the research and development at major corporations is now moving away from artificial-intelligence research, in which the computer functions separately, to research on the human–computer symbiosis, including information environments that augment human bodily perception and create "virtual realities."[9] While we may legitimately inquire into the power wielded by computers independently of humans, the existential–ontological question really cuts in a different and deeper direction than does AI. As we now live and work with computers in our writing, building, banking, drawing, and so forth, how does our reality change? As Heidegger might put it: What is the meaning of this intimate connection of Being with computers? When he pondered technology as our destiny, Heidegger seemed to have had something in mind more intimidating than an external challenge to our dignity as human beings. What Heidegger saw was something even more sinister than a revolt of the machines.

What Heidegger called "the essence of technology" infiltrates human existence more intimately than anything humans could create. The danger of technology lies in the transformation of the human being, by which human actions and aspirations are fundamentally distorted. Not that machines can run amok, or even that we might misunderstand ourselves through a faulty comparison with machines. Instead, technology enters the inmost recesses of human existence, transforming the way we know and think and will. Technology is, in essence, a mode of human existence, and we could not appreciate its mental infiltration until the computer became a major cultural phenomenon.

Already in 1957 Heidegger noticed the drive for technological mastery pushing into the human interior where thought and reality meet in language. In his essay on Hebel, he wrote:

■ The language machine regulates and adjusts in advance the mode of our possible usage of language through mechanical energies and functions. The language machine is—and above all, is still becoming—one way in which modern technology controls the mode and the world of language as such. Meanwhile, the impression is still maintained that man is the master of the language machine. But the truth of the matter might well be that the language machine takes language into its management and thus masters the essence of the human being.[10]

What did Heidegger mean when he referred to the "language machine" (*Sprachmaschine*)? He did not say "computer"—

Heidegger and McLuhan: The Computer as Component

the only computers around then were huge mainframes like the UNIVAC that filled several rooms and performed only numerical calculations. Could we, twenty-five years later, translate what Heidegger meant by using the English term *computer* instead of *language machine*?

In the early 1980s, I ran into the meaning of language machine just as I was finishing the translation of another book by Heidegger, *The Metaphysical Foundations of Logic*. The translation required a lot of detailed organization, since the main body of the text includes extensive citations in Latin, French, and Greek. As a scholarly publication, the text had not only to render but also to preserve many of the references in their original languages. The translation required laborious cross-referencing with other texts and other English translations. Index cards and cut-and-paste scraps swamped my kitchen table. Perseverance was all. The work took me more than two years. Then, just as I finished typing the third and final draft of the translation, I discovered the language machine, the connection between Heidegger and computers. Not long after mailing the final draft of the translation, I installed my own personal computer for word processing. Imagine my mixed feelings when I came to realize that the two years of labor on the translation would have amounted to no more than one year if I had used a computer to handle the text and references. The meaning of language machine began to take shape in my mind.

The Computer as Component

Soon after trading in my electric typewriter for a portable computer in 1983, I came to believe that the machine in my hands was indeed the language machine of Heidegger's speculations. The "language machine" was Heidegger's groping term for the incipient phenomenon of word processing. Of course, word processing did not exist in Heidegger's lifetime, at least not as a cultural phenomenon. It existed only in the dreams of inventors like Doug Engelbart and Ted Nelson. Although he did not see the word processor, Heidegger did have a keen eye for the philosophical implications in the shift of writing technologies. He saw in writing technology a clue to the human relationship to language and to our awareness as beings embodied in the world:

> Not by chance does modern man write "with" the typewriter and "dictates"—the same word as "to invent creatively" [*Dichten*]—"into" the machine. This "history" of the kinds of writing is at the same time one of the major reasons for the increasing destruction of the word. The word no longer passes through the hand as it writes and acts authentically, but through the mechanized pressure of the hand. The typewriter snatches script from the essential realm of the hand—and this means the hand is removed from the essential realm of the word. The word becomes something "typed." Nevertheless, mechanical script does have its own, limited importance in which mechanized script serves as a mere transcription for preserving handwriting or in which typewritten script substitutes for "print." When typewriters first became prevalent, a personal letter typed on a machine was regarded as a lapse of manners or as an insult. Today, handwritten letters slow down rapid reading and are therefore regarded as old-fashioned and undesirable. Mechanized writing deprives the hand of dignity in the realm of the written word and degrades the word to a mere means for the traffic of communication.[11]

Heidegger focuses on the increasing typification brought about by modern rationalist models of standardized intelligibility, models that underscore the advantages of repetition and instant recognition.

Heidegger's criticisms of the typewriter are somewhat off the mark now that the personal computer has replaced the mechanical typewriter. Unlike the typewriter, the word processor guides the hand into a nonmechanical process. The fingers on the keyboard might just as well be a voice that activates the information device, for the computer removes the writing activity from script and mechanical imprints. Word processing can also have a graphic interface that brings the hand back to bodily gestures like pointing and moving things around with a graphic pointing device or mouse. The actions are carried out by an already typified, digitized element. Unlike the typewriter, the computer does not simply replace direct hand movements with the industrial–mechanical action of springs, pulleys, and levers. The information environment allows gestures to work in ways that leave behind the industrial machine with its cumbersome but efficient mediation of human energy and attention. The electronic element shifts the quality of action to another level. The formulation of ideas on a word processor can establish impersonality while achieving a directness and flexibility undreamed of with the typewriter.

Heidegger sensed the power of the machine as an agent for changing our relationship to the word. In fact, the word processor changes our relationship to written language at least as much as the printing press does. Nor can scholarship go unchanged. Heidegger correctly feared that electronic digital text might absorb his own work. In 1967, he feared that a rising tide of information might soon obliterate his own writings: "Maybe history and tradition will fit smoothly into the information retrieval systems which will serve as resource for the inevitable planning needs of a cybernetically organized mankind. The question is whether thinking too will end in the business of information processing."[12]

If it has already transformed the epistemic stance of the natural sciences, the computer is transforming the humanities as well. The word processor is the calculator of the humanist, giving its users the power to manipulate written language in new ways. Just as the printing press altered culture and scholarship soon after its invention, so too the computer automates the composition, storage, and transmission of written words. And if the computer affects all written communication, will it not in turn affect the way in which we regard and use language in general—not only when we sit at the word processor but also, by aftereffect, whenever we speak and listen, perhaps even whenever we think?

Computer technology is so flexible and adaptable to our thought processes that we soon consider it less an external tool and more a second skin or mental prosthesis. Once acclimated to the technology, we play it much as a musician plays an instrument, identifying with it, becoming one with it. Writing on the language machine produces a new kind of writing and thinking. At our fingertips is the calculating machine dreamed of by Pascal and Leibniz, the fathers of modern metaphysics, but now this calculator operates on our language as we spontaneously produce it.

Heidegger sensed that the language machine belongs to our destiny. What did he mean when he said the language machine would "take language into its management and master the essence of the human being"? Was he simply reacting to change? Should we place him historically among the reactionaries of his time?[13] I think not. Political terms of reaction or progress are too crude here. Heidegger's statement invites us to insight, not political agendas. He was meditating on a technology still in the bud. Now that this technology is blossoming, we need to consider what he was getting at. Neither Luddite nor technophobe, Heidegger resisted every attempt to categorize his views as either optimistic or pessimistic. Whether the glass was half-empty or half-full, Heidegger was interested in the substance of its contents. He was a soft determinist, accepting destiny while studying the different ways of absorbing its impact. In this respect, he resembled the Canadian philosopher of communication Marshall McLuhan.

McLuhan and Computers

Like McLuhan, Heidegger believed that he had grasped something unique and essential about the twentieth century. Both Heidegger and McLuhan felt an inner relationship to their epoch. Each believed that he was interpreting a destiny that the next generation would receive, and each believed that the legacy of his reflections on technology was far more important than his own personal value judgments about technology. McLuhan wrote that he held back his own value judgments from the public because they create "a smog in our culture." He wrote: "I have tried to avoid making personal value judgments about these processes [of technological transformation] since they seem far too important and too large in scope to deserve a merely private opinion."[14] Similarly, Heidegger held back statements of personal values from his philosophy, whether statements of self-justifications or of a moral agenda.

The point was to reflect on the radical shifts brought about by an unprecedented development.

Both Heidegger and McLuhan saw intimate connections between information technology and the way the mind works. If Heidegger is the father of information anxiety, McLuhan is the child of the television medium of the 1960s. What synchronized their visions is the crucial role that technology plays in defining reality, in operating as an invisible backdrop within which the content or entities of the world appear. Behind the visible entities of the world, McLuhan glimpsed a hidden backdrop:

■ To say that any technology or extension of man creates a new environment is a much better way of saying the medium is the message. Moreover, this environment is always "invisible" and its content is always the old technology. The old technology is altered considerably by the enveloping action of the new technology.[15]

For Heidegger, likewise, the question of technology was not an ontic one, not one about the proliferation of devices or even about the possible supremacy of the machine over human beings. His ontological question touches the world, the clearing or backdrop against which things appear. Ontology has to do with our understanding of the being of things, not with things as such. The ontological question probes the invisible background. As McLuhan saw it, "The content of the new environment is always the old one. The content is greatly transformed by the new technology. . . . Today the environment itself becomes the artifact."[16] Technology would not sweep the older things away but would transform them while placing them before us as though nothing had changed. Similarly, according to Heidegger, the future takes up the past while making it present, and the environment we live in quickly becomes an artifact in the omnivorous future of the technological system.

McLuhan helps us understand what the computer does specifically as a language machine, as a component of human knowledge. Both McLuhan and Heidegger considered the most awesome power of technology to reside in its newly achieved intimacy with language. McLuhan noted with approval Heidegger's treatment of language as a transcendental aspect of Being:

> The alphabet and kindred gimmicks have long served man as a subliminal source of philosophical and religious assumptions. Certainly Martin Heidegger would seem to be on better ground [than Kant was in assuming Euclidian space to be an a priori] in using the totality of language itself as philosophical datum. For there, at least in non-literate periods, will be the ratio among all the senses. . . . An enthusiasm for Heidegger's excellent linguistics could easily stem from naive immersion in the metaphysical organicism of our electronic milieu. . . . There is nothing good or bad about print but the unconsciousness of the effect of any force is a disaster, especially a force that we have made ourselves.[17]

McLuhan suggests that Heidegger's ideas have a greater appeal to a culture organized electronically because such a culture has already left behind the detached, linear, individualistic mentality of literate or print cultures. He agrees with Heidegger in asserting that language technology belongs to us more essentially than does any other tool. When a technology touches our language, it touches us where we live.

How can we philosophically reflect on the word processor? How can we get beyond the vague general talk about the dangers of the calculative mentality? McLuhan's work can help track the impact of word-processing technology more specifically and clearly. But for me it was not McLuhan but an illustrious student of his, Walter J. Ong, who provided a more precise conceptual angle from which I could better see the language machine. For specific insight into the way that the word processor alters our thought processes and even our sense of reality, I found help in the writings of Ong, who treats the psychodynamics—the shifts in mentality—that occur in Western history as new technologies for language storage come into prominence.

Ong traces two major shifts in knowledge storage: the oral-to-literate and the chirographic-to-print shifts. The first occurred when the culture moved from a predominantly oral-based society to a society increasingly based on the written word. The second shift moved from handwritten (chirographic) texts to the more widely disseminated, mechanically produced printed books. With more detail and coherence than his mentor McLuhan, Ong traced these shifts in writing technologies as they affected human awareness and in turn influenced interactive epistemology (knowledge as it occurs in relation to tools and to other persons). Unlike an absolute stance, this epistemological approach takes seriously the changes that mark the history of human knowledge. The studies by Ong and Eric Havelock (*Preface to Plato*) provide concrete material for distinguishing different historical epochs by their characteristic ways of symbolizing, storing, and transmitting truths. The patterns of psychic transformation they trace dovetail nicely with Heidegger's history of being.

According to Heidegger, we notice the eclipse of the truth of being occurring already in Plato's metaphysics. Once the truth of being becomes equated with the light of unchanging intelligibility, the nature of truth shifts to the ability of statements to reflect or refer reliably to entities. With the steadiness of propositional truth comes the tendency to relate to being as a type, a form, or an anticipated shape. With being as a steady form, entities gain their reality through their being typified. Already in Plato we see the seeds of the Western drive to standardize things, to find what is dependable and typical in them. Truth as the disclosure process, as the play of revealing/concealing disappears behind the scene in which the conscious mind grasps bright objects apprehended as clear, unwavering, rational forms. As humans develop the ability to typify and apprehend formal realities, the loss of truth as emergent disclosure goes unnoticed. All is light and form. Nothing hides behind the truth of beings. But this "nothing" finally makes an appearance after the whole world has become a rigid grid of standardized forms and shapes conceived and engineered by humans. As the wasteland grows, we see the devastation of our fully explicit truths. We see that there is, must be, more. The hidden extra cannot be consciously produced. Only by seeing the limits of standardization can we begin to respond to it. We have to realize that each advance in typifying and standardizing things also implies a trade-off. When we first reach forward and grasp things, we see only the benefits of our standardization, only the positive side of greater clarity and utility. It is difficult to accept the paradox that no matter how alluring, every gain in fixed intelligibility brings with it a corresponding loss of vivacity. Because we are finite, every gain we make also implies a lost possibility. The loss is especially devastating to those living in the technological world, for here they enjoy everything conveniently at their disposal—everything, that is, except the playful process of discovery itself.

The McLuhan-inspired theory of cultural transformation brings out even more sharply the impact of the word processor. But this theory lacks a poignant sense of loss or a feel for the trade-offs in finite historical transformations. Ong's version of cultural transformation has about it something of a grand Christian optimism, seeing in the global network of electronic radio, television, and film a way of reintegrating a fallen, fragmented humanity, creating a closer community. For Ong, the shift from a predominantly oral culture to a literate culture shattered the original tribal unity. In bringing about greater individualism and fostering the logical faculties, literacy cut into the psychic roots of belonging and severed the attachment to immediate interpersonal presence. The print culture even further reinforced literacy, spreading it ever more widely, lifting individualism to unprecedented heights. Then, in Hegelian fashion, Ong sees the electronic media sublating the earlier oppositions, the oral and the literate, so that electronics achieves an encompassing synthesis. Electronic visuals, supported by voices, re-creates human presence and reunites the individuated members of the community. Underneath, however, the electronic images still depend on the reading of scripts, prepared messages, and a print-informed society. So the electronic media preserve individual literacy while surpassing it. Because of his hopeful Hegelian dialectic, Ong omits the critical evaluation that can take place only in the existential moment. Whereas McLuhan remained publicly silent on the adverse effects of the new media, Ong appears to have absorbed criticism in a larger pic-

Heidegger and McLuhan: The Computer as Component

ture based on the Christian narrative of Garden→Fall→
Paradise Regained.

Heidegger, on the contrary, reminds us of the inevitable
trade-offs in history. His philosophy does in fact proceed from
the Hegelian sweep of historical epochs, but it denies the possibility of an integrative summation from one absolute standpoint. History is a series of ambiguous gains bringing hidden
losses. The series of epochs that makes up the history of reality (*Seinsgeschichte*) expands or contracts with different hermeneutic projects but never permits a single cumulative
narrative. Each moment of historical transformation brings
a challenge of interpreting the losses and gains, the trade-offs
in historical drift. The drift of history allows no safe haven
from which to assess and collect strictly positive values once
and for all.

In our era, Heidegger's notion of the intrinsic trade-offs
of history can spark a critical analysis of computerized writing. Existential criticism can investigate the implications of a
specific technology in all its ambiguity. Because it accepts
historical drift, existential criticism proceeds without possess-
ing a total picture of the whither and wherefore, without accepting the picture promoted by either technological utopians
or dystopians. There is no need to enforce a closure of pro
or con, wholesale acceptance or rejection. While recognizing
the computer as a component in our knowledge process, we
can attend to what happens to us as we collaborate with technology. Because human history is a path of self-awareness,
as we deepen our understanding of computer interaction, we
will also increase our self-understanding.

Notes

1. Ralph Holibaugh, director of the Olin and Chalmers libraries
   at Kenyon College, in *The Kenyon College Annual Report 1988–
   90*, p. 5.
2. Martin Heidegger, Preface to *Wegmarken* (Frankfurt: Klostermann,
   1967), p. i. [Author's translation]
3. Don Ihde, for one, sees science as merging with its instruments

in *Technology and the Lifeworld: From Garden to Earth* (Bloomington: Indiana University Press, 1990).

4. Rainer Bast and Heinrich Delfosse, *Handbuch zum Textstudium von Martin Heideggers "Sein und Zeit"* (Stuttgart: Frommann-Holzboog, 1979).

5. Hubert Dreyfus, *What Computers Can't Do: The Limits of Artificial Intelligence*, rev. ed. (New York: Harper Colophon, 1979).

6. Hubert Dreyfus, *Mind over Machine: The Power of Human Intuition and Expertise in the Era of the Computer* (New York: Free Press, 1985).

7. The history of this chess match appears in Howard Rheingold, *Tools for Thought: The People and Ideas Behind the Next Computer Revolution* (New York: Simon and Schuster, 1985), pp. 161–62. Dreyfus explains what he takes to be the point of the match in *Mind over Machine* (p. 112).

8. Martin Heidegger, "Platons Lehre von der Wahrheit," in *Wegmarken*, pp. 109–44. This was a lecture series given in 1930 and 1931.

9. See Michael Benedikt, ed., *Cyberspace: First Steps* (Cambridge, Mass.: MIT Press, 1991). The term *cyberspace* originated with William Gibson, who uses science fiction to explore the symbiotic connection of humans and computers.

10. Martin Heidegger, *Hebel—Der Hausfreund* (Pfullingen: Günther Neske, 1957); reprinted as "Hebel—Friend of the House," trans. Bruce Foltz and Michael Heim, in *Contemporary German Philosophy*, ed. Darrel E. Christensen (University Park: Pennsylvania State University Press, 1983), vol. 3, pp. 89–101.

11. Martin Heidegger, *Parmenides*, vol. 54 of *Gesamtausgabe* (Frankfurt: Klostermann, 1982), pp. 119–19. [Author's translation] These were lectures given in the winter of 1942/1943. In this passage, Heidegger is commenting on the ancient Greek notion of "action" (*pragma*).

12. Heidegger, Preface to *Wegmarken*, p. ii. [Author's translation]

13. A recent study that locates Heidegger's theory of technology in the cultural reaction of the Weimar Republic is Michael Zimmerman, *Heidegger's Confrontation with Modernity: Technology, Politics, Art* (Bloomington: Indiana University Press, 1990).

14. Marshall McLuhan to Jonathan Miller, April 1970, in *The Letters of Marshall McLuhan*, comp. and ed. Matie Molinaro, Corinne McLuhan, and William Toye (New York: Oxford University Press, 1987), p. 405. In a letter to Jonathan Miller, April 1970, McLuhan wrote: "I take it that you understand that I have never expressed any preferences or values since *The Mechanical Bride*. Value judgments create smog in our culture and distract attention from processes. My personal bias is entirely pro-print and all

of its effects" (ibid.). In other places, McLuhan was not so open about his stance. In writing to Eric Havelock, May 1970, for instance, he noted:

> My own studies of the effects of technology on human psyche and society have inclined people to regard me as the enemy of the things I describe. I feel a bit like the man who turns in a fire alarm only to be charged with arson. I have tried to avoid making personal value judgments about these processes since they seem far too important and too large in scope to deserve a merely private opinion. (ibid., p. 406)

15. McLuhan to John Culkin, September 1964, in ibid., p. 309.
16. McLuhan to Buckminster Fuller, September 1964, in ibid., p. 309.
17. H. Marshall McLuhan, *The Gutenberg Galaxy: The Making of Typographic Man* (Toronto: University of Toronto Press, 1962), p. 66, in a section entitled "Heidegger surf-boards along on the electronic wave as triumphantly as Descartes rode the mechanical wave."

■ 6

FROM INTERFACE
TO CYBERSPACE

When IBM personal computers first landed on office desks
in late August 1981, some cheered but most squirmed. A January 1983 issue of *Time* conveyed the general technoanxiety
in its cover story: Instead of the cover portrait of its customary
"Man of the Year," *Time* displayed a portrait of the "Machine
of the Year," a desktop computer that was the machine of the
decade, if not indeed of the century.

The portrait showed a bleak scene sculpted by George
Segal: standing on an old-fashioned wooden table, a personal
computer displays colorful charts and words. In front of the
computer, a chalk-white, life-cast, male figure sits on a
wooden chair. He slumps slightly, hands immobile on the
knees, and stares passively at the screen. Nearby, another
desktop computer occupies a smaller table where a female
life-cast sits in a wicker chair. She relaxes cross-legged, coffee
cup in hand, turning her line of vision away from the screen.
Both cast figures stand out white against a black background
where a single large white-framed window looks out on inky
blackness. The computers and furniture beam bright primary
colors, while the surrounding atmosphere seems claustro-
phobic. Mostly colorless and nearly empty, the room suggests
an oppressive banality in which humans are spiritless peripherals of their information devices.

The initial computer phobia of the 1980s has ebbed
away. In its place is a fear that computers may have simply

added another layer of stress. Instead of simplifying or solving problems, computers may have brought on addictive overwork and automated joblessness. Rather than increase productivity, computers in business seem to have multiplied activities without ensuring substantive results. The vague anxiety about the "Machine of the Year" has given way to a strong suspicion that we are long on machines but short on accomplishments. Technoanxiety has turned into a skepticism laced with the sobering conviction that there is no turning back. The anxiety that haunted the 1980s was like the day before the wedding. Nobody knew what the future would hold, only that the future would be unimaginably different from the past.

Now we are wedded to machines, for better or worse. The first phase of our marriage was the appliance. Whether in the kitchen or on the freeway, much of our time is spent directing, tending, or waiting for machines. So closely do we work with devices that we seldom notice them—until they break down. Machines are no longer merely machines but have become electromechanical appliances. Applied technology fills our lives with familiar routines. The routines tick according to the watches and clocks that govern the tempo of our lives and coordinate our whereabouts. We are constantly reminded that we inhabit a grid of social time–space. Timepieces band our wrists like tags on the legs of homing pigeons.

Modern life with its appliances is a fairly recent experiment. Our distant ancestors put little stock in machines, considering artificial devices mere novelties or toys at best. Centuries ago in ancient China, Taoist sages were openly hostile to machines. One day a young man observed an old sage fetching water from the village well. The old man lowered a wooden bucket on a rope and pulled the water up hand over hand. The youth disappeared and returned with a wooden pulley. He approached the old man and showed him how the device works. "See, you put your rope around the wheel and draw the water up by cranking on the handle." The old man replied, "If I use a device like this, my mind will think itself clever. With a cunning mind, I will no longer put my heart into what I am doing. Soon my wrist alone will do the work, turning the handle. If my heart and whole body is not in my work, my work will become joyless. When my work is joyless, how do you think the water will taste?"

We in the West do not accept the Taoist dilemma between clever-minded alienation and wholehearted brute labor. We opt for something in between. We have converted occasional devices into appliances, integral parts of the daily routine. Sure, we have our Luddites and machine manglers, but, as a rule, we want to have our Zen and do motorcycle maintenance, too. Our technology summons our passion. We get involved in things in and through the technology. In this respect, we resemble the citizens of Samuel Delaney's novel *Nova* who are outfitted with sockets, one in each wrist and one at the base of the spine. The sockets provide the psychoneural energy for running everything from vacuum cleaners to factories or starships. Instead of removing people from their work, our technology connects us to our work, putting us directly into our activities. Devices attach to every aspect of life, creating a technological culture. Our marriage to technology embraces production, transportation, and communication.

When technology becomes a interconnected system, at some point we sit down and address the system itself; we program it. When a technology covers a number of particular tasks, linking them into a system, then the next step is to gain a vantage point where we can obtain information about the system and can call programs for the system to run. No longer a set of discrete machines, our appliances communicate with one another and function through information. A cybernetic infrastructure coordinates instruments that measure everything from weather and traffic flow to banking transactions. We feed input into the system, which then constantly feeds information back to us. Our selves plus the machines constitute a feedback loop. Information brings into existence the information age, and the holy shrine of our era becomes the computer screen. Then we have the next stage of our marriage to technology. The newlyweds begin to influence each other. We go from appliance to interface.

What is an interface? Today the term *interface* works as a noun or a verb. Its meaning ranges from computer cables to television screens and it describes everything from personal meetings to corporate financial mergers. *Interface* is a buzzword. Buzzwords cue conversations, directing our attention momentarily. Occasionally a buzzword strikes a deeper resonance: the word buzzes, we push on it, and suddenly a magic door swings open to who we are and where we stand in history. Such buzzwords are keywords. Keywords are not just fuzzy metaphors or poetic symbols of an epoch, like "the computer age" or "the nuclear age." Rather, keywords apply analogously across the entire spectrum of cultural life. They apply accurately and in detail to many aspects of our lives— each aspect differently. Keywords cut across our whole cultural world, and *interface* is a keyword.

The meaning of interface runs from economics to metaphysics. In the 1990s, American economists recognize how crucial interface is, and they understand the word in a material sense, as a viewing screen; for them, interface is a hardware product that must be developed and brought to market. The latest hardware is high-definition television (HDTV), a video interface with sharp images. American industry, many economists say, will stand or fall with high-definition television. Economists and business consultants have registered their belief in the importance of HDTV for American economic interests. Economic well-being depends on how soon American industry can produce these flat-screen displays for home video, for other countries are competing for the video market.

The importance of the state-of-the-art interface goes beyond economic competition and includes national defense. The U.S. military depends on the latest video displays. The Department of Defense needs high-resolution monitors for its future line of helicopters, planes, ships, and tanks. Pilots and soldiers need high-definition screens in their training simulators. High-tech warfare calls for split-second coordination between humans and machines. A nanosecond delay or a slight distortion in visual information can spell disaster.

Medicine, too, needs high-definition displays. Medical diagnosis is leaving behind the shadowy x-ray. Nuclear magnetic resonance (NMR) reveals far more visual information than any photograph could provide. First, a magnetic field of gamma rays measures organs and even molecular structures. Taking these data as input, the computer then reconstructs entire cross sections of the patient's body. The visual output must be sharp enough to show a lesion one millimeter long or detailed enough for doctors to distinguish a left or right cusp of the aortic valve. Medical imaging differs from interpreting photographs, as the new healing technology requires less intuitive guesswork based on what the human eye can barely make out. Healing belongs to the interface.

Our link to the solar system, the stars, and other galaxies comes through the interface. All areas of the universe come into sharper focus as scientists zoom in on sources of astronomical data. In the late summer of 1989, scientists at the Jet Propulsion Laboratory in Pasadena, California, sat before their video screens viewing Triton, a moon of Neptune, as the *Voyager 2* space probe sent radio signals back to earth. Through representations and simulations, we contact the world we know and even the limits of what we know.

The interface is taking to the freeway, too, where the automobile is the grandfather of applied technology. In the near future, motorists will have on-board computer displays of maps, locations, and detailed directions to any destination. Traffic information will appear on the car windshield, projected by virtual images above the driver's line of sight. The auto interface will provide navigational readouts, warnings about blind spots, and television reception. Pilots have long used the heads-up displays on jet aircraft, sometimes even landing in foggy weather by viewing a televised simulation of the runway. Driving any vehicle may soon mean using a video interface to enhance vision. This enhancement the developers call *reality augmentation*, for it superimposes information on a direct reality percept. The combination "augmented reality" is a step toward breaking through the interface and inhabiting an electronic realm where reality and symbolized reality constitute a third entity: virtual reality.

What, then, is an interface? An interface occurs where two or more information sources come face-to-face. A human user connects with the system, and the computer becomes interactive. Tools, by contrast, establish no such connection.

We use tools, picking them up or putting them down. They do not adjust to our purposes, except in the most primitive physical sense. The wrench fits into my hand and allows wider or narrower settings. The electric screwdriver offers various speeds. Still, the wrench does not become a screwdriver, nor does the screwdriver help me remove a nut. A piece of software, on the contrary, permits me to make any number of tools for different jobs. My word processor, for instance, allows me to do many things: outline ideas, search and compare texts, write and read messages, check spelling, index books or notes, do arithmetic calculations, or program any sequence of operations by means of macros. The software interface is a two-way street where computers enhance and modify my thinking power.

Interface means more than video hardware, more than a screen we look at. Interface refers also to software or to the way we actively alter the computer's operations and consequently alter the world controlled by the computer. Interface denotes a contact point where software links the human user to computer processors. This is the mysterious, nonmaterial point where electronic signals become information. It is our interaction with software that creates an interface. Interface means the human being is wired up. Conversely, technology incorporates humans.

In ancient times, the term *interface* sparked awe and mystery. The archaic Greeks spoke reverently of *prosopon*, or a face facing another face. Two opposite faces make up a mutual relationship. One face reacts to the other, and the other face reacts to the other's reaction, and the other reacts to that reaction, and so on ad infinitum. The relationship then lives on as a third thing or state of being. The ancient term *prosopon* once glowed with mystic wonder. The same word later helped Christians describe their Trinitarian Godhead. The Father and the Son subsist together as an interface or distinct spirit. The ancient word suggests a spiritual interaction between eternity and time.

In the information age, a mystic glow surrounds the term *cyberspace*. Every type of interface mentioned earlier forms a window or doorway into cyberspace. Cyberspace suggests a computerized dimension where we move information about and where we find our way around data. Cyberspace renders a represented or artificial world, a world made up of the information that our systems produce and that we feed back into the system. Just as a chessboard sets up the checkered game space of its own world of rooks and knights, pawns and bishops, so too the computer interface holds its field of moves, hierarchy of files, places to go, and relative distances between points of interest. We inhabit cyberspace when we feel ourselves moving through the interface into a relatively independent world with its own dimensions and rules. The more we habituate ourselves to an interface, the more we live in cyberspace, in what William Gibson calls the "consensual hallucination."[1]

As the interface shades into cyberspace, the dark atmosphere of George Segal's sculpture seeps in. This interface brings with it a troubling ambiguity. The term *interface* originated with the mundane hardware adapter plugs used to connect electronic circuits. Then it came to mean the video hardware used to peer into the system. Finally, it denotes the human connection with machines, even the human entry into a self-contained cyberspace. In one sense, interface indicates computer peripherals and video screens; in another, it indicates human activity connected by video to data. The double meaning makes us pause: How peripheral is the human? How much of the system do we own when we enter the door of cyberspace? Where are we when software architects shape the datascape into endless mazes of light attracting us like moths to a flame? The very idea of an interface points to an impending future that may be our fate, even fatal. The cover of *Time* floats like an afterimage.

Cyberspace can cast a spell of passivity on our lives. We talk to the system, telling it what to do, but the system's language and processes come to govern our psychology. We begin as voyeurs and end by abandoning our identity to the fascinating systems we tend. The tasks beckoning us to the network make us forget our elemental loss in the process. We look through the interface unaware as we peer through an electronic framework where our symbols—words, data, simula-

From Interface to Cyberspace

tions—come under precise control, where things appear with startling clarity. So entrancing are these symbols that we forget ourselves, forget where we are. We forget ourselves as we evolve into our fabricated worlds. With our faces up against it, the interface is hard to see. Because information technology fits our minds, it is the hardest of all to think about. Nothing is closer to us. We can miss it as easily as we overlook a pair of eyeglasses on the bridge of the nose or a contact lens on the cornea.

Gibson refers to cyberspace as "an infinite cage."[2] We can travel endlessly in cyberspace, without limits, for cyberspace is electronic, and electronically we can represent not only the actual physical universe but also possible and imagined worlds. But to a finite incarnate being, such an infinity constitutes a cage, a confinement to a nonphysical secondary realm.

Virtual-reality systems can use cyberspace to represent physical space, even to the point that we feel telepresent in a transmitted scene, whether Mars or the deep ocean. But the data building the cyberworld pull the user away from the internal bioenergies that run our primary body. The interface belongs to minds that love to represent. As we sit before the interface, like the life-cast figures of the Segal sculpture, we reinforce the representing mind-set. We experience a world under control, even while observing how out of control the world may be at any particular moment. No matter what is represented, the interface supplies a shape and a form. It guides and even warps our visual imagination. People growing up today with films and television see things in video format. Our dreams and imaginings often occur to us as if we were watching television. We find it difficult to become aware of our own internal states without the objective representations of the interface.

We do not even realize when we are trapped in our minds and cybersystems. The basic world we incarnate gradually is lost in our attention to the cognitive and imagined worlds. To offset the loss, we need to preserve what remains of the wisdom of Far Eastern philosophy, particularly Taoism with its profound system of internal body awareness. Taoists discovered how to enhance human vitality by focusing the mind's internal attention on the body. Acupuncture, Tai Chi, and yogic healing are based on Taoist tradition.

Eastern philosophy has always sought to unify the mind and the body, to harmonize outer and inner. Various exercises, from Yoga to Zen and Tai Chi, bring the mind/body into a harmonious unity. From the Eastern viewpoint, the body houses private, subtle energy, and this subtle energy remains internal. No representation adequately fixes it in images. The strength of the mind/body grows when the mind's attention synchronizes with the internal energy. Meditators and martial artists join awareness to the breathing process, because breathing links the subconscious autonomic nervous system to conscious life. One sign of the disruption of the smooth flow of internal energy is holding the breath. Holding the breath or shallow breathing is the respiratory equivalent of the stare we typically use at the interface. And when we stare, we usually hold our breath.

The deepest peril of the interface is that we may lose touch with our inner states. By inner states, I do not mean anything arcane. The Taoists urge us to contact our inner physical organs, to "see" our liver, "smell" our lungs, and "taste" our heart. By this they mean something quite simple. They mean not to lose the acute sensitivity to our bodies, the simplest kinds of awareness like kinesthetic body movement, organic discomfort, and propriosensory activities like breathing, balance, and shifting weight. The loss of such simple inner states may seem trivial. Taken as a whole, however, this awareness constitutes the background for the psychic life of the individual. "The body is the temple of the spirit."

One far-seeing inventor of virtual-reality systems, Myron Krueger, has dedicated his life to bringing full freedom of body movement to the interface. His work has yet to blossom, as most developers produce systems that shackle the body with goggles, gloves, and datasuits. But even if we learn to interact with computerized objects and remain unencumbered by oppressive VR systems, will we maintain enough power of additional awareness needed for both primary and secondary worlds—at the same time? At the very least, we will need to retrain our powers of attention, just for the sake of long-term mental and physical health.

The growing interface exacerbates a tension already built into modern life. In his preface to a book by F. M. Alexander, John Dewey saw an "internecine warfare at the heart of our civilization between the functions of the brain and the nervous system on one side and the functions of digestion, circulation, respiration and the muscular system on the other." Dewey regarded this conflict as "a perilous affair." A half-century before the computer interface, Dewey observed: "If our habitual judgments of ourselves are warped, because they are based on vitiated sense material, then the more complex the social conditions under which we live, the more disastrous must be the outcome. Every additional complication of outward instrumentalities is likely to be a step nearer destruction."[3]

In the 1960s, Jim Morrison saw the danger to sensibility in *The Lords and the New Creatures*, in which he warned: "There may be a time when we'll attend Weather Theatre to recall the sensation of rain."[4] Back then, Morrison could not know that the Weather Theater will soon be everywhere and that we will need lessons in recalling why we love the sensation of the rain.

Notes

1. William Gibson, *Neuromancer* (New York: Ace Books, 1984), p. 51; Gibson, *Count Zero* (New York: Ace Books, 1986), p. 38.
2. William Gibson, *Mona Lisa Overdrive* (New York: Bantam Books, 1988), p. 49.
3. John Dewey, Preface to Frederick Matthias Alexander, *The Resurretion of the Body: The Writings of F. M. Alexander*, ed. Edward Maisel (New York: University Books, 1969), p. 169.
4. Jim Morrison, *The Lords and the New Creatures* (New York: Simon and Schuster, 1969), p. 64.

■ 7

THE EROTIC ONTOLOGY
OF CYBERSPACE

Cyberspace is more than a breakthrough in electronic media or in computer interface design. With its virtual environments and simulated worlds, cyberspace is a metaphysical laboratory, a tool for examining our very sense of reality.

When designing virtual worlds, we face a series of reality questions. How, for instance, should users appear to themselves in a virtual world? Should they appear to themselves in cyberspace as one set of objects among others, as third-person bodies that users can inspect with detachment? Or should users feel themselves to be headless fields of awareness, similar to our phenomenological experience? Should causality underpin the cyberworld so that an injury inflicted on the user's cyberbody likewise somehow damages the user's physical body? And who should make the ongoing design decisions? If the people who make simulations inevitably incorporate their own perceptions and beliefs, loading cyberspace with their prejudices as well as their insights, who should build the cyberworld? Should multiple users at any point be free to shape the qualities and dimensions of cyber entities? Should artistic users roam freely, programming and directing their own unique cyber cinemas that provide escape from the mundane world? Or does fantasy cease where the economics of the virtual workplace begins? But why be satisfied with a single virtual world? Why not several? Must we pledge allegiance to a single reality? Perhaps worlds should be layered like onion skins, realities within realities, or be loosely linked like neighborhoods, permitting free aesthetic pleasure to coexist with the task-oriented business world. Does the meaning of "reality"—and the keen existential edge of experience—weaken as it stretches over many virtual worlds?

Important as these questions are, they do not address the ontology of cyberspace itself, the question of what it means to *be* in a virtual world, whether one's own or another's world. They do not probe the reality status of our metaphysical tools or tell us why we invent virtual worlds. They are silent about the essence or soul of cyberspace. How does the metaphysical laboratory fit into human inquiry as a whole? What status do electronic worlds have within the entire range of human experience? What perils haunt the metaphysical origins of cyberspace?

In what follows, I explore the philosophical significance of cyberspace. I want to show the ontological origin from which cyber entities arise and then indicate the trajectory they seem to be on. The ontological question, as I see it, requires a two-pronged answer. We need to give an account of (1) the way entities exist within cyberspace and (2) the ontological status of cyberspace—the construct, the phenomenon—itself. The way in which we understand the ontological structure of cyberspace will determine how realities can exist within it. But the structure of cyberspace becomes clear only once we appreciate the distinctive way in which things appear within it. So we must begin with the entities we experience within the computerized environment.

My approach to cyberspace passes first through the ancient idealism of Plato and moves onward through the modern metaphysics of Leibniz. By connecting with intellectual precedents and prototypes, we can enrich our self-understanding and make cyberspace function as a more useful metaphysical laboratory.

Our Marriage to Technology

The phenomenal reality of cyber entities exists within a more general fascination with technology, and the fascination with technology is akin to aesthetic fascination. We love the simple, clear-cut linear surfaces that computers generate. We love the way that computers reduce complexity and ambiguity, capturing things in a digital network, clothing them in beaming colors, and girding them with precise geometrical structures. We are enamored of the possibility of controlling all human knowledge. The appeal of seeing society's data structures in cyberspace—if we begin with William Gibson's vision—is like the appeal of seeing the Los Angeles metropolis in the dark at five thousand feet: a great warmth of powerful, incandescent blue and green embers with red stripes that beckons the traveler to come down from the cool darkness. We are the moths attracted to flames, and frightened by them too, for there may be no home behind the lights, no secure abode behind the vast glowing structures. There are only the fiery objects of dream and longing.

Our love affair with computers, computer graphics, and computer networks runs deeper than aesthetic fascination and deeper than the play of the senses. We are searching for a home for the mind and heart. Our fascination with computers is more erotic than sensuous, more spiritual than utilitarian. Eros, as the ancient Greeks understood, springs from a feeling of insufficiency or inadequacy. Whereas the aesthete feels drawn to casual play and dalliance, the erotic lover reaches out to a fulfillment far beyond aesthetic detachment.

The computer's allure is more than utilitarian or aesthetic; it is erotic. Instead of a refreshing play with surfaces, as with toys or amusements, our affair with information machines announces a symbiotic relationship and ultimately a mental marriage to technology. Rightly perceived, the atmosphere of cyberspace carries the scent that once surrounded Wisdom. The world rendered as pure information not only fascinates our eyes and minds, but also captures our hearts. We feel augmented and empowered. Our hearts beat in the machines. This is Eros.

Cyberspace entities belong to a broad cultural phenomenon of the last third of the twentieth century: the phenomenon of computerization. Something becomes a phenomenon when it arrests and holds the attention of a civilization. Only then

The Erotic Ontology of Cyberspace

does our shared language articulate the presence of the thing so that it can appear in its steady identity as the moving stream of history.

Because we are immersed in everyday phenomena, however, we usually miss their overall momentum and cannot see where they are going or even what they truly are. A writer like William Gibson helps us grasp what is phenomenal in current culture because he captures the forward movement of our attention and shows us the future as it projects its claim back into our present. Of all writers, Gibson most clearly reveals the intrinsic allure of computerized entities, and his books—*Neuromancer*, *Count Zero*, and *Mona Lisa Overdrive*—point to the near-future, phenomenal reality of cyberspace. Indeed, Gibson coined the word *cyberspace*.

The Romance of *Neuromancer*

For Gibson, cyber entities appear under the sign of Eros. The fictional characters of *Neuromancer* experience the computer matrix—cyberspace—as a place of rapture and erotic intensity, of powerful desire and even self-submission. In the matrix, things attain a supervivid hyper-reality. Ordinary experience seems dull and unreal by comparison. Case, the data wizard of *Neuromancer*, awakens to an obsessive Eros that drives him back again and again to the information network:

> A year [in Japan] and he still dreamed of cyberspace, hope fading nightly. . . . [S]till he'd see the matrix in his sleep, bright lattices of logic unfolding across that colorless void. . . . [H]e was no [longer] console man, no cyberspace cowboy. . . . But the dreams came on in the Japanese night like livewire voodoo, and he'd cry for it, cry in his sleep, and wake alone in the dark, curled in his capsule in some coffin hotel, his hands clawed into the bedslab, . . . trying to reach the console that wasn't there.[1]

The sixteenth-century Spanish mystics John of the Cross and Teresa of Avila used a similar point of reference. Seeking words to connote the taste of spiritual divinity, they reached for the language of sexual ecstasy. They wrote of the breathless

union of meditation in terms of the ecstatic blackout of consciousness, the *llama de amor viva* piercing the interior center
of the soul like a white-hot arrow, the *cauterio suave* searing
through the dreams of the dark night of the soul. Similarly, the
intensity of Gibson's cyberspace inevitably conjures up the
reference to orgasm, and vice versa:

■ Now she straddled him again, took his hand, and closed it over
her, his thumb along the cleft of her buttocks, his fingers spread
across the labia. As she began to lower herself, the images came
pulsing back, the faces, fragments of neon arriving and reced-
ing. She slid down around him and his back arched con-
vulsively. She rode him that way, impaling herself, slipping
down on him again and again, until they both had come, his
orgasm flaring blue in a timeless space, a vastness like the matrix, where the faces were shredded and blown away down
hurricane corridors, and her inner thighs were strong and wet
against his hips.[2]

But the orgasmic connection does not mean that Eros's going
toward cyberspace entities terminates in a merely physiologi-
cal or psychological reflex. Eros goes beyond private, subjec-
tive fantasies. Cyber Eros stems ultimately from the ontological
drive highlighted long ago by Plato. Platonic metaphysics
helps clarify the link between Eros and computerized entities.

In her speech in Plato's *Symposium*, Diotima, the
priestess of love, teaches a doctrine of the escalating spirituality of the erotic drive. She tracks the intensity of Eros con-
tinuously from bodily attraction all the way to the mental
attention of mathematics and beyond. The outer reaches of
the biological sex drive, she explains to Socrates, extend to
the mental realm where we continually seek to expand our
knowledge.

On the primal level, Eros is a drive to extend our finite
being, to prolong something of our physical selves beyond
our mortal existence. But Eros does not stop with the drive
for physical extension. We seek to extend ourselves and to
heighten the intensity of our lives in general through Eros. The
psyche longs to perpetuate itself and to conceive offspring,
and this it can do, in a transposed sense, by conceiving ideas

The Erotic Ontology of Cyberspace

and nurturing awareness in the minds of others as well as our own. The psyche develops consciousness by formalizing perceptions and stabilizing experiences through clearly defined entities. But Eros motivates humans to see more and to know more deeply. So, according to Plato, the fully explicit formalized identities of which we are conscious help us maintain life in a "solid state," thereby keeping perishability and impermanence at bay.

Only a short philosophical step separates this Platonic notion of knowledge from the matrix of cyberspace entities. (The word *matrix*, of course, stems from the Latin for "mother," the generative–erotic origin). A short step in fundamental assumptions, however, can take centuries, especially if the step needs hardware support. The hardware for implementing Platonically formalized knowledge took centuries. Underneath, though, runs an ontological continuity, connecting the Platonic knowledge of ideal forms to the information systems of the matrix. Both approaches to cognition first extend and then renounce the physical embodiment of knowledge. In both, Eros inspires humans to outrun the drag of the "meat"—the flesh—by attaching human attention to what formally attracts the mind. As Platonists and Gnostics down through the ages have insisted, Eros guides us to Logos.

The erotic drive, however, as Plato saw it, needs education to attain its fulfillment. Left on its own, Eros naturally goes astray on any number of tangents, most of which come from sensory stimuli. In the *Republic*, Plato tells the well-known story of the Cave in which people caught in the prison of everyday life learn to love the fleeting, shadowy illusions projected on the walls of the dungeon of the flesh. With their attention forcibly fixed on the shadowy moving images cast by a flickering physical fire, the prisoners passively take sensory objects to be the highest and most interesting realities. Only later, when the prisoners manage to get free of their corporeal shackles, do they ascend to the realm of active thought, where they enjoy the shockingly clear vision of real things, things present not to the physical eyes but to the mind's eye. Only by actively processing things through mental logic, according to Plato, do we move into the upper air of reliable truth, which is also a lofty realm of intellectual beauty stripped of the imprecise impressions of the senses. Thus the liberation from the Cave requires a reeducation of human desires and interests. It entails a realization that what attracts us in the sensory world is no more than an outer projection of ideas we can find within us. Education must redirect desire toward the formally defined, logical aspects of things. Properly trained, love guides the mind to the well-formed, mental aspects of things.

Cyberspace is Platonism as a working product. The cybernaut seated before us, strapped into sensory-input devices, appears to be, and is indeed, lost to this world. Suspended in computer space, the cybernaut leaves the prison of the body and emerges in a world of digital sensation.

This Platonism is thoroughly modern, however. Instead of emerging in a sensationless world of pure concepts, the cybernaut moves among entities that are well formed in a special sense. The spatial objects of cyberspace proceed from the constructs of Platonic imagination not in the same sense that perfect solids or ideal numbers are Platonic constructs, but in the sense that inFORMation in cyberspace inherits the beauty of Platonic FORMS. The computer recycles ancient Platonism by injecting the ideal content of cognition with empirical specifics. Computerized representation of knowledge, then, is not the direct mental insight fostered by Platonism. The computer clothes the details of empirical experience so that they seem to share the ideality of the stable knowledge of the Forms. The mathematical machine uses a digital mold to reconstitute the mass of empirical material so that human consciousness can enjoy an integrity in the empirical data that would never have been possible before computers. The notion of ideal Forms in early Platonism has the allure of a perfect dream. But the ancient dream remained airy, a landscape of genera and generalities, until the hardware of information retrieval came to support the mind's quest for knowledge. Now, with the support of the electronic matrix, the dream can incorporate the smallest details of here-and-now existence. With an electronic infrastructure, the dream of perfect FORMS becomes the dream of inFORMation.

Filtered through the computer matrix, all reality becomes patterns of information. When reality becomes indistinguishable from information, then even Eros fits the schemes of binary communication. Bodily sex appears to be no more than an exchange of signal blips on the genetic corporeal network. Further, the erotic–generative source of formal idealism becomes subject to the laws of information management. Just as the later Taoists of ancient China created a yin–yang cosmology that encompassed sex, cooking, weather, painting, architecture, martial arts, and the like, so too the computer culture interprets all knowable reality as transmissible information. The conclusion of *Neuromancer* shows us the transformation of sex and personality into the language of information:

■ There was a strength that ran in her, . . . [s]omething he'd found and lost so many times. It belonged, he knew—he remembered—as she pulled him down, to the meat, the flesh the cowboys mocked. It was a vast thing, beyond knowing, a sea of information coded in spiral and pheromone, infinite intricacy that only the body, in its strong blind way, could ever read.

. . . [H]e broke [the zipper], some tiny metal part shooting off against the wall as salt-rotten cloth gave, and then he was in her, effecting the transmission of the old message. Here, even here, in a place he knew for what it was, a coded model of some stranger's memory, the drive held.

She shuddered against him as the stick caught fire, a leaping flare that threw their locked shadows across the bunker wall.[3]

The dumb meat once kept sex private, an inner sanctum, an opaque, silent, unknowable mystery. The sexual body held its genetic information with the strength of a blind, unwavering impulse. What is translucent you can manipulate, you can see. What stays opaque you cannot scrutinize and manipulate. It is an alien presence. The meat we either dismiss or come up against; we cannot ignore it. It remains something to encounter. Yet here, in *Neuromancer*, the protagonist, Case, makes love to a sexual body named Linda. Who is this Linda?

Gibson raises the deepest ontological question of cyberspace by suggesting that the Neuromancer master-computer *simulates* the body and personality of Case's beloved. A simulated, embodied personality provokes the sexual encounter. Why? Perhaps because the cyberspace system, which depends on the physical space of bodies for its initial impetus, now seeks to undermine the separate existence of human bodies that make it dependent and secondary. The ultimate revenge of the information system comes when the system absorbs the very identity of the human personality, absorbing the opacity of the body, grinding the meat into information, and deriding erotic life by reducing it to a transparent play of puppets. In an ontological turnabout, the computer counterfeits the silent and private body from which mental life originated. The machinate mind disdainfully mocks the meat. Information digests even the secret recesses of the caress. In its computerized version, Platonic Eros becomes a master of artificial intelligence, CYBEROS, the controller, the Neuromancer.

The Inner Structure of Cyberspace

Aware of the phenomenal reality of cyber entities, we can now appreciate the backdrop that is cyberspace itself. We can sense a distant source radiating an all-embracing power. For the creation of computerized entities taps into the most powerful of our psychobiological urges. Yet so far, this account of the distant source as Eros tells only half the story. For although Platonism provides the psychic makeup for cyberspace entities, only modern philosophy shows us the structure of cyberspace itself.

In its early phases—from roughly 400 B.C. to A.D. 1600—Platonism exclusively addressed the speculative intellect, advancing a verbal–mental intellectuality over physical actuality. Later, Renaissance and modern Platonists gradually injected new features into the model of intelligence. The modern Platonists opened up the gates of verbal–spiritual understanding to concrete experiments set in empirical space and time. The new model of intelligence included the evidence of repeatable experience and the gritty details of experiment. For the first time, Platonism would have to absorb real space and real time into the objects of its contemplation.

The early Platonic model of intelligence considered space to be a mere receptacle for the purely intelligible entities subsisting as ideal forms. Time and space were refractive errors that rippled and distorted the mental scene of perfect unchanging realities. The bouncing rubber ball was in reality a round object, which was in reality a sphere, which was in reality a set of concentric circles, which could be analyzed with the precision of Euclidian geometry. Such a view of intelligence passed to modern Platonists, and they had to revise the classical assumptions. Thinkers and mathematicians would no longer stare at the sky of unchanging ideals. By applying mathematics to empirical experiment, science would absorb physical movement in space/time through the calculus. Mathematics transformed the intelligent observer from a contemplator to a calculator. But as long as the calculator depended on feeble human memory and scattered printed materials, a gap would still stretch between the longing and the satisfaction of knowledge. To close the gap, a computational engine was needed.

Before engineering an appropriate machine, the cyberspace project needed a new logic and a new metaphysics. The new logic and metaphysics of modernity came largely from the work of Gottfried Leibniz. In many ways, the later philosophies of Kant, Schopenhauer, Nietzsche, and Heidegger took their bearings from Leibniz.

As Leibniz worked out the modern Idealist epistemology, he was also experimenting with protocomputers. Pascal's calculator had been no more than an adding machine; Leibniz went further and produced a mechanical calculator that could also, by using stepped wheels, multiply and divide. The basic Leibnizian design became the blueprint for all commercial calculators until the electronics revolution of the 1970s. Leibniz, therefore, is one of the essential philosophical guides to the inner structure of cyberspace. His logic, metaphysics, and notion of representational symbols show us the hidden underpinnings of cyberspace. At the same time, his monadological metaphysics alerts us to the paradoxes that are likely to engulf cyberspace's future inhabitants.

Leibniz's Electric Language

Leibniz was the first to conceive of an "electric language," a set of symbols engineered for manipulation at the speed of thought. His *De arte combinatoria* (1666) outlines a language that became the historical foundation of contemporary symbolic logic. Leibniz's general outlook on language also became the ideological basis for computer-mediated telecommunications. A modern Platonist, Leibniz dreamed of the matrix.

The language that Leibniz outlined is an ideographic system of signs that can be manipulated to produce logical deductions without recourse to natural language. The signs represent primitive ideas gleaned from prior analysis. Once broken down into primitives and represented by stipulated signs, the component ideas can be paired and recombined to fashion novel configurations. In this way, Leibniz sought to mechanize the production of new ideas. As he described it, the encyclopedic collection and definition of primitive ideas would require the coordinated efforts of learned scholars from all parts of the civilized world. The royal academies that Leibniz promoted were the group nodes for an international republic of letters, a universal network for problem solving.

Leibniz believed all problems to be, in principle, soluble. The first step was to create a universal medium in which conflicting ideas could coexist and interrelate. A universal language would make it possible to translate all human notions and disagreements into the same set of symbols. His universal character set, *characteristica universalis*, rests on a binary logic, one quite unlike natural discourse in that it is neither restricted by material content nor embodied in vocalized sound. Contentless and silent, the binary language can transform every significant statement into the terms of a logical calculus, a system for proving argumentative patterns valid or invalid, or at least for connecting them in a homogeneous matrix. Through the common binary language, discordant ways of thinking can exist under a single roof. Disagreements in attitude or belief, once translated into matching symbols, can later yield to operations for ensuring logical consistency. To the partisans of dispute, Leibniz would say, "Let us upload this into our common system, then let us sit down and calculate." A single system would encompass all the combinations and permutations of human thought. Leibniz longed for his symbols to foster unified scientific research throughout the civilized world. The universal calculus would compile all human culture, bringing every natural language into a single shared database.

Leibniz's binary logic, disembodied and devoid of material content, depends on an artificial language remote from the words, letters, and utterances of everyday discourse. This logic treats reasoning as nothing more than a combining of signs, as a calculus. Like mathematics, the Leibnizian symbols erase the distance between the signifiers and the signified, between the thought seeking to express and the expression. No gap remains between symbol and meaning. Given the right motor, the Leibnizian symbolic logic—as developed later by George Boole, Bertrand Russell, and Alfred North Whitehead and then applied to electronic switching circuitry by Shannon—can function at the speed of thought. At such high speed, the felt semantic space closes between thought, language, and the thing expressed. Centuries later, John von Neumann applied a version of Leibniz's binary logic when building the first computers at Princeton.

In his search for a universal language of the matrix, Leibniz to some extent continued a premodern, medieval tradition. For behind his ideal language stands a premodern model of human intelligence. The medieval Scholastics held that human thinking, in its pure or ideal form, is more or less identical with logical reasoning. Reasoning functions along the lines of a superhuman model who remains unaffected by the vagaries of feelings and spatiotemporal experience. Human knowledge imitates a Being who knows things perfectly and knows them in their deductive connections. The omniscient Being transcends finite beings. Finite beings go slowly, one step at a time, seeing only moment by moment what is happening. On the path of life, a finite being cannot see clearly the things that remain behind on the path or the things that are going to happen after the next step. A divine mind, on the contrary, oversees the whole path. God sees all the trails below, inspecting at a single glance every step traveled, what has happened, and even what will happen on all possible paths below. God views things from the perspective of the mountaintop of eternity.

Human knowledge, thought Leibniz, should emulate this *visio dei*, this omniscient intuitive cognition of the deity. Human knowledge strives to know the way that a divine or an infinite Being knows things. No temporal unfolding, no linear steps, no delays limit God's knowledge of things. The temporal simultaneity, the all-at-once-ness of God's knowledge serves as a model for human knowledge in the modern world as projected by the work of Leibniz. What better way, then, to emulate God's knowledge than to generate a virtual world constituted by bits of information? To such a cyberworld human beings could enjoy a God-like instant access. But if knowledge is power, who would handle the controls that govern every single particle of existence?

The power of Leibniz's modern logic made traditional logic seem puny and inefficient by comparison. For centuries, Aristotle's logic had been taught in the schools. Logic traditionally evaluated the steps of finite human thought, valid or invalid, as they occur in arguments in natural language. Traditional logic stayed close to spoken natural language. When modern logic absorbed the steps of Aristotle's logic into its system of symbols, modern logic became a network of symbols that could apply equally to electronic switching circuits as to arguments in natural language. Just as non-Euclidian geometry can set up axioms that defy the domain of real circles (physical figures), so too modern logic freed itself of any naturally given syntax. The universal logical calculus could govern computer circuits.

Leibniz's "electric language" operates by emulating the divine intelligence. God's knowledge has the simultaneity of all-at-once-ness, and so in order to achieve a divine access to things, the global matrix functions like a net to trap all language in an eternal present. Because access need not be linear, cyberspace does not, in principle, require a jump from one location to another. Science fiction writers have often imagined what it would be like to experience traveling at the speed of light, and one writer, Isaac Asimov, described such travel as a "jump through hyperspace." When his fictional space ship hits the speed of light, Asimov says that the ship makes a special kind of leap. At that speed, it is impossible to trace the discrete points of the distance traversed. In the novel *The Naked Sun*, Asimov depicts movement in hyperspace:

▪ There was a queer momentary sensation of being turned inside out. It lasted an instant and Baley knew it was a Jump, that oddly incomprehensible, almost mystical, momentary transition through hyperspace that transferred a ship and all it contained from one point in space to another, light years away. Another lapse of time and another Jump, still another lapse, still another Jump.[4]

Like the fictional hyperspace, cyberspace unsettles the felt logical tracking of the human mind. Cyberspace is the perfect computer environment for accessing hypertext if we include all human perceptions as the "letters" of the "text." In both hyperspace and hypertext, linear perception loses track of the series of discernible movements. With hypertext, we connect things at the speed of a flash of intuition. The interaction with hypertext resembles movement beyond the speed of light. Hypertext reading and writing supports the intuitive leap over the traditional step-by-step logical chain. The jump, not the step, is the characteristic movement in hypertext.

As the environment for sensory hypertext, cyberspace feels like transportation through a frictionless, timeless medium. There is no jump because everything exists, implicitly if not actually, all at once. To understand this lightning speed and its perils for finite beings, we must look again at the metaphysics of Leibniz.

Monads Do Have Terminals

Leibniz called his metaphysics a *monadology*, a theory of reality describing a system of "monads." From our perspective, the monadology conceptually describes the nature of beings who are capable of supporting a computer matrix. The monadology can suggest how cyberspace fits into the larger world of networked, computerized beings.

The term *monadology* comes from the Greek *monas*, as in "monastic," "monk," and "monopoly." It refers to a certain kind of aloneness, a solitude in which each being pursues its appetites in isolation from all other beings, which also are solitary. The monad exists as an independent point of vital willpower, a surging drive to achieve its own goals according to its own internal dictates. Because they are a sheer, vital thrust, the monads do not have inert spatial dimensions but produce space as a by-product of their activity. Monads are nonphysical, psychical substances whose forceful life is an immanent activity. For monads, there is no outer world to access, no larger, broader vision. What the monads see are the projections of their own appetites and their own ideas. In Leibniz's succinct phrase: "Monads have no windows."

Monads may have no windows, but they do have terminals. The mental life of the monad—and the monad has no other life—is a procession of internal representations. Leibniz's German calls these representations *Vorstellungen*, from *vor* (in front of) and *stellen* (to place). Realities are representations continually placed in front of the viewing apparatus of the monad, but placed in such a way that the system interprets or represents what is being pictured. The monad sees the pictures of things and knows only what can be pictured. The monad knows through the interface. The interface represents things, simulates them, and preserves them in a format that the monad can manipulate in any number of ways. The monad keeps the presence of things on tap, as it were, making them instantly available and disposable, so that the presence of things is represented or "canned." From the vantage point of physical phenomenal beings, the monad undergoes a surrogate experience. Yet the monad does more than think about or imagine things at the interface. The monad senses things, sees them and hears them as perceptions. But the perceptions of phenomenal entities do not occur in real physical space because no substances other than monads really exist. Whereas the interface with things vastly expands the monad's perceptual and cognitive powers, the things at the interface are simulations and representations.

Yet Leibniz's monadology speaks of monads in the plural. For a network to exist, more than one being must exist; otherwise, nothing is there to be networked. But how can monads coordinate or agree on anything at all, given their isolated nature? Do they even care if other monads exist? Leibniz tells us that each monad represents within itself the entire universe. Like Indra's Net, each monad mirrors the whole world. Each monad represents the universe in concentrated form, making within itself a *mundus concentratus*. Each microcosm contains the macrocosm. As such, the monad reflects the universe in a living mirror, making it a *miroir actif indivisible*, whose appetites drive it to represent everything to itself—everything, that is, mediated by its mental activity. Since each unit represents everything, each unit contains all the other units, containing them as represented. No direct physical contact passes between the willful mental units. Monads never meet face-to-face.

Although the monads represent the same universe, each one sees it differently. The differences in perception come from differences in perspective. These different perspectives arise not from different physical positions in space—the monads are not physical, and physical space is a by-product of mental perception—but from the varying degrees of clarity and intensity in each monad's mental landscape. The appetitive impulses in each monad highlight different things in the sequence of representational experience. Their different impulses constantly shift the scenes they see. Monads run different software.

Still, there exists, according to the monadology, one actual universe. Despite their ultimately solitary character, the monads belong to a single world. The harmony of all the entities in the world comes from the one underlying operating system. Although no unit directly contacts other units, each unit exists in synchronous time in the same reality. All their representations are coordinated through the supervisory role of the Central Infinite Monad, traditionally known as God. The Central Infinite Monad, we could say, is the Central System Operator (sysop), who harmonizes all the finite monadic units. The Central System Monad is the only being that exists with absolute necessity. Without a sysop, no one could get on line to reality. Thanks to the Central System Monad, each individual monad lives out its separate life according to the dictates of its own willful nature while still harmonizing with all the other monads on line.

Paradoxes in the Cultural Terrain of Cyberspace

Leibniz's monadological metaphysics brings out certain aspects of the erotic ontology of cyberspace. Although the monadology does not actually describe computerized space, of course, it does suggest some of the inner tendencies of computerized space. These tendencies are inherent in the structure of cyberspace and therefore affect the broader realities in which the matrix exists. Some paradoxes crop up. The monadological metaphysics shows us a cultural topography riddled with deep inconsistencies.

Cyberspace supplants physical space. We see this happening already in the familiar cyberspace of on-line communication—telephone, e-mail, newsgroups, and so forth. When on line, we break free, like the monads, from bodily existence. Telecommunication offers an unrestricted freedom of expression and personal contact, with far less hierarchy and formality than are found in the primary social world. Isolation persists as a major problem of contemporary urban society, and I mean spiritual isolation, the kind that plagues individuals even on crowded city streets. With the telephone and television, the computer network can function as a countermeasure. The computer network appears as a godsend in providing forums for people to gather in surprisingly personal proximity—especially considering today's limited bandwidths—without the physical limitations of geography, time zones, or conspicuous social status. For many, networks and bulletin boards act as computer antidotes to the atomism of society. They assemble the monads. They function as social nodes for fostering those fluid and multiple elective affinities that everyday urban life seldom, in fact, supports.

Unfortunately, what technology gives with one hand, it often takes away with the other. Technology increasingly eliminates direct human interdependence. While our devices give us greater personal autonomy, at the same time they disrupt the familiar networks of direct association. Because our machines automate much of our labor, we have less to do with one another. Association becomes a conscious act of will. Voluntary associations operate with less spontaneity than do those having sprouted serendipitously. Because machines provide us with the power to flit about the universe, our communities grow more fragile, airy, and ephemeral even as our connections multiply.

Being a *body* constitutes the principle behind our separateness from one another and behind our personal presence. Our bodily existence stands at the forefront of our personal identity and individuality. Both law and morality recognize the physical body as something of a fence, an absolute boundary, establishing and protecting our privacy. Now the computer network simply brackets the physical presence of the participants, by either omitting or simulating corporeal immediacy. In one sense, this frees us from the restrictions imposed by our physical identity. We are more equal on the net because we can either ignore or create the body that appears in cyberspace. But in another sense, the quality of the human encounter narrows. The secondary or stand-in body reveals only as much of ourselves as we mentally wish to reveal. Bodily contact becomes optional; you need never stand face-to-face with other members of the virtual community. You can live your own separate existence without ever physically meeting another person. Computers may at first liberate societies through increased communication and may even foment revolutions (I am thinking of the computer printouts in Tiananmen Square during the 1989 prodemocracy uprisings in China). They have, however, another side, a dark side.

The darker side hides a sinister melding of human and machine. The cyborg, or cybernetic organism, implies that the conscious mind steers—the meaning of the Greek *kybernetes*—our organic life. Organic life energy ceases to initiate our mental gestures. Can we ever be fully present when we live through a surrogate body standing in for us? The stand-in self lacks the vulnerability and fragility of our primary identity. The stand-in self can never fully represent us. The more we mistake the cyberbodies for ourselves, the more the machine twists ourselves into the prostheses we are wearing.

Gibson's fiction inspired the creation of role-playing games for young people. One of these games in the cybertech genre, *The View from the Edge: The Cyberpunk Handbook*, portrays the visage of humanity twisted to fit the shapes of the computer prosthesis. The body becomes literally "meat" for the implantation of information devices. The computer plugs directly into the bones of the wrist or skull and taps into major nerve trunks so that the chips can send and receive neural signals. As the game book wryly states:

■ Some will put an interface plug at the temples (a "plug head"), just behind the ears (called a "frankenstein") or in the back of the head (a "puppethead"). Some cover them with inlaid silver or gold caps, others with wristwarmers. Once again, a matter of style. Each time you add a cybernetic enhancement, there's a corresponding loss of humanity. But it's not nice, simple and linear. Different people react differently to the cyborging process. Therefore, your humanity loss is based on the throw of random dice value for each enhancement. This is important, because it means that sheer luck could put you over the line before you know it. Walk carefully. Guard your mind.[5]

At the computer interface, the spirit migrates from the body to a world of total representation. Information and images float through the Platonic mind without a grounding in bodily experience. You can lose your humanity at the throw of the dice.

The Erotic Ontology of Cyberspace

Gibson highlights this essentially Gnostic aspect of cyber-tech culture when he describes the computer addict who despairs at no longer being able to enter the computer matrix: "For Case, who'd lived for the bodiless exultation of cyberspace, it was the Fall. In the bars he'd frequented as a cowboy hotshot, the elite stance involved a certain relaxed contempt for the flesh. The body was meat. Case fell into the prison of his own flesh."[6] The surrogate life in cyberspace makes flesh feel like a prison, a fall from grace, a descent into a dark confusing reality. From the pit of life in the body, the virtual life looks like the virtuous life. Gibson evokes the Gnostic–Platonic–Manichean contempt for earthy, earthly existence.

Today's computer communication cuts the physical face out of the communication process. Computers stick the windows of the soul behind monitors, headsets, and datasuits. Even video conferencing adds only a simulation of face-to-face meeting, only a representation or an appearance of real meeting. The living, nonrepresentable face is the primal source of responsibility, the direct, warm link between private bodies. Without directly meeting others physically, our ethics languishes. Face-to-face communication, the fleshly bond between people, supports a long-term warmth and loyalty, a sense of obligation for which the computer-mediated communities have not yet been tested. Computer networks offer a certain sense of belonging, to be sure, but the sense of belonging circulates primarily among a special group of pioneers. How long and how deep are the personal relationships that develop outside embodied presence? The face is the primal interface, more basic than any machine mediation. The physical eyes are the windows that establish the neighborhood of trust. Without the direct experience of the human face, ethical awareness shrinks and rudeness enters. Examples abound. John Coates, spokesperson for the WELL in northern California says: "Some people just lose good manners on line. You can really feel insulated and protected from people if you're not looking at them—nobody can take a swing at you. On occasion, we've stepped in to request more diplomacy. One time we had to ask someone to go away."[7]

At the far end of distrust lies computer crime. The machine interface may amplify an amoral indifference to human relationships. Computers often eliminate the need to respond directly to what takes place between humans. People do not just observe one another, but become "lurkers." Without direct human presence, participation becomes optional. Electronic life converts primary bodily presence into telepresence, introducing a remove between represented presences. True, in bodily life we often play at altering our identity with different clothing, masks, and nicknames, but electronics installs the illusion that we are "having it both ways," keeping our distance while "putting ourselves on the line." On-line existence is intrinsically ambiguous, like the purchased passion of the customers in the House of Blue Lights in Gibson's *Burning Chrome*: "The customers are torn between needing someone and wanting to be alone at the same time, which has probably always been the name of that particular game, even before we had the neuroelectronics to enable them to have it both ways."[8] As the expanding global network permits the passage of bodily representations, "having it both ways" may reduce trust and spread cynical anomie.

A loss of innocence therefore accompanies an expanding network. As the on-line culture grows geographically, the sense of community diminishes. Shareware worked well in the early days of computers, and so did open bulletin boards. When the size of the user base increased, however, the spirit of community diminished, and the villains began appearing, some introducing viruses. Hackers invisibly reformatted hard disks, and shareware software writers moved to the commercial world. When we speak of a global village, we should keep in mind that every village makes villains, and when civilization reaches a certain degree of density, the barbaric tribes return, from within. Tribes shun their independent thinkers and punish individuality. A global international village, fed by accelerated competition and driven by information, may be host to an unprecedented barbarism. Gibson's vision of cyberspace works like a mental aphrodisiac, but it turns the living environment—electronic and real—into a harsh, nightmarish jungle. This jungle is more than a mere cyberpunk

The Erotic Ontology of Cyberspace

affectation, a matter of aestheticizing grit or conflict or rejection. It may also be an accurate vision of the intrinsic energies released in a cyberized society.

An artificial information jungle already spreads out over the world, duplicating with its virtual vastness the scattered geography of the actual world. The matrix already multiplies confusion, and future cyberspace may not simply reproduce a more efficient version of traditional information. The new information networks resemble the modern megalopolis, often described as a concrete jungle (New York) or a sprawl (Los Angeles). A maze of activities and hidden byways snakes around with no apparent center. Architecturally, the network sprawl suggests the absence of a philosophical or religious absolute. Traditional publishing resembles a medieval European city, with the center of all activity, the cathedral or church spire, guiding and gathering all the communal directions and pathways. The steeple visibly radiates like a hub, drawing the inhabitants into a unity and measuring the other buildings on a central model. Traditionally, the long-involved process of choosing which texts to print or which movies or television shows to produce serves a similar function. The book industry, for instance, provides readers with various cues for evaluating information. The publishers legitimize printed information by giving clues that affect the reader's willingness to engage in reading the book. Editorial attention, packaging endorsements by professionals or colleagues, book design, and materials all add to the value of the publisher's imprint. Communication in contemporary cyberspace lacks the formal clues. In their place are private recommendations or just blind luck. The electronic world, unlike the traditional book industry, does not protect its readers or travelers by following rules that set up certain expectations. Already, in the electric element, the need for stable channels of content and reliable processes of choice grows urgent.

If cyberspace unfolds like existing large-scale media, we might expect a debasement of discriminating attention. If the economics of marketing forces the matrix to hold the attention of a critical mass of the population, we might expect a flashy liveliness and a flimsy currency to replace depth of content.

Sustained attention will give way to fast-paced cuts. One British humanist spoke of the HISTORY forum on Bitnet in the following terms: "The HISTORY network has no view of what it exists for, and of late has become a sort of bar-room courthouse for pseudo-historical discussion on a range of currently topical events. It really is, as Glasgow soccer players are often called, a waste of space." Cyberspace without carefully laid channels of choice may become a waste of space.

The Underlying Fault

Finally, on-line freedom seems paradoxical. If the drive to construct cyber entities comes from Eros in the Platonic sense, and if the structure of cyberspace follows the model of Leibniz's computer God, then cyberspace rests dangerously on an underlying fault of paradox. Remove the hidden recesses, the lure of the unknown, and you also destroy the erotic urge to uncover and reach further; you destroy the source of yearning. Set up a synthetic reality, place yourself in a computer-simulated environment, and you undermine the human craving to penetrate what radically eludes you, what is novel and unpredictable. The computer God's-eye view robs you of your freedom to be fully human. Knowing that the computer God already knows every nook and cranny deprives you of your freedom to search and discover.

Even though the computer God's-eye view remains closed to the human agents in cyberspace, they will know that such a view exists. Computerized reality synthesizes everything through calculation, and nothing exists in the synthetic world that is not literally numbered and counted. Here Gibson's protagonist gets a brief glimpse of this superhuman, or inhuman, omniscience:

▪ Case's consciousness divided like beads of mercury, arcing above an endless beach the color of the dark silver clouds. His vision was spherical, as though a single retina lined the inner surface of a globe that contained all things, if all things could be counted.

And here things could be counted, each one. He knew

the number of grains of sand in the construct of the beach (a number coded in a mathematical system that existed nowhere outside the mind that was Neuromancer). He knew the number of yellow food packets in the canisters in the bunker (four hundred and seven). He knew the number of brass teeth in the left half of the open zipper of the salt-crusted leather jacket that Linda Lee wore as she trudged along the sunset beach, swinging a stick of driftwood in her hand (two hundred and two).[9]

The erotic lover reels under the burden of omniscience: "If all things could be counted . . ." Can the beloved remain the beloved when she is fully known, when she is fully exposed to the analysis and synthesis of binary construction? Can we be touched or surprised—deeply astonished—by a synthetic reality, or will it always remain a magic trick, an illusory prestidigitation?

With the thrill of free access to unlimited corridors of information comes the complementary threat of total organization. Beneath the artificial harmony lies the possibility of surveillance by the all-knowing Central System Monad. The absolute sysop wields invisible power over all members of the network. The infinite CSM holds the key for monitoring, censoring, or rerouting any piece of information or any phenomenal presence on the network. The integrative nature of the computer shows up today in the ability of the CSM to read, delete, or alter private e-mail on any computer-mediated system. Those who hold the keys to the system, technically and economically, have access to anything on the system. The CSM will most likely place a top priority on maintaining and securing its power. While matrix users feel geographical and intellectual distances melt away, the price they pay is their ability to initiate uncontrolled and unsupervised activity.

According to Leibniz's monadology, the physical space perceived by the monads comes as an inessential by-product of experience. Spatiotemporal experience goes back to the limitations of the fuzzy finite monad minds, their inability to grasp the true roots of their existence. From the perspective of eternity, the monads exist by rational law and make no unprescribed movements. Whatever movement or change they make disappears in the lightning speed of God's absolute cognition. The flesh, Leibniz maintained, introduces a cognitive fuzziness. For the Platonic imagination, this fuzzy incarnate world dims the light of intelligence.

Yet the erotic ontology of cyberspace contradicts this preference for disembodied intelligibility. If I am right about the erotic basis of cyberspace, then the surrogate body undoes its genesis, contradicts its nature. The ideal of the simultaneous all-at-once-ness of computerized information access undermines any world that is worth knowing. The fleshly world is worth knowing for its distances and its hidden horizons. Thankfully, the Central System Monad never gets beyond the terminals into the physical richness of this world. Fortunately, here in the broader world, we still need eyes, fingers, mice, modems, and phone lines.

Gibson leaves us the image of a human group that instinctively keeps its distance from the computer matrix. These are the Zionites, the religiously tribal folk who prefer music to computers and intuitive loyalties to calculation. The Zionites constitute a human remnant in the environmental desolation of *Neuromancer:*

■ Case didn't understand the Zionites.
    . . . The Zionites always touched you when they were talking, hands on your shoulder. He [Case] didn't like that. . . .
    "Try it," Case said [holding out the electrodes of the cyberspace deck].
    [The Zionite Aerol] took the band, put it on, and Case adjusted the trodes. He closed his eyes. Case hit the power stud. Aerol shuddered. Case jacked him back out. "What did you see, man?"
    "Babylon," Aerol said, sadly, handing him the trodes and kicking off down the corridor.[10]

As we suit up for the exciting future in cyberspace, we must not lose touch with the Zionites, the body people who remain rooted in the energies of the earth. They will nudge us out of our heady reverie in this new layer of reality. They will remind us of the living genesis of cyberspace, of the heartbeat behind the laboratory, of the love that still sprouts amid the

The Erotic Ontology of Cyberspace

broken slag and the rusty shells of oil refineries "under the poisoned silver sky."

Notes

1. William Gibson, *Neuromancer* (New York: Ace Books, 1984), pp. 4–5.
2. Ibid., p. 33.
3. Ibid., pp. 239–40.
4. Isaac Asimov, *The Naked Sun* (New York: Ballantine, 1957), p. 16.
5. Mike Pondsmith, *The View from the Edge: The Cyberpunk Handbook* (Berkeley, Calif.: R. Talsorian Games, 1988), pp. 20–22.
6. Gibson, *Neuromancer*, p. 6.
7. Quoted in Steve Rosenthal, "Turn On, Dial Up, Tune In," *Electric Word*, November–December 1989, p. 35.
8. William Gibson, *Burning Chrome* (New York: Ace Books, 1987), p. 191.
9. Gibson, *Neuromancer*, p. 258.
10. Ibid., p. 106.

■ 8

THE ESSENCE OF VR

What is virtual reality?

A simple enough question.

We might answer: "Here, try this arcade game. It's from the Virtuality series created by Jonathan Waldern. Just put on the helmet and the datagloves, grab the control stick, and enter a world of computer animation. You turn your head and you see a three-dimensional, 360-degree, color landscape. The other players see you appear as an animated character. And lurking around somewhere will be the other animated warriors who will hunt you down. Aim, press the button, and destroy them before they destroy you. Give it a few minutes and you'll get a feel for the game, how to move about, how to be part of a virtual world. That's virtual reality!"

Suppose the sample experience does not satisfy the questioner. Our questioner has already played the Virtuality game. Suppose the question is about virtual reality in general.

Reach for a dictionary. *Webster's* states:

Virtual: "being in essence or effect though not formally
recognized or admitted"
Reality: "a real event, entity, or state of affairs"

We paste the two together and read: "Virtual reality is an event or entity that is real in effect but not in fact."

Not terribly enlightening. You don't learn nuclear physics from dictionaries. We need insight, not word usage.

The dictionary definition does, however, suggest something about VR. There is a sense in which any simulation makes something seem real that in fact is not. The Virtuality game combines head-tracking device, glove, and computer animation to create the "effect" on our senses of "entities" moving at us that are "not in fact real."

But what makes VR distinctive? "What's so special," our questioner might ask, "about these computer-animated monsters? I've seen them before on television and in film. Why call them 'virtual realities'?"

The questioner seeks not information, but clarification.

Pointing to the helmet and gloves, we insist: "Doesn't this feel a lot different from watching TV? Here you can inter- act with the animated creatures. You shoot them down or hide from them or dodge their ray guns. And they interact with you. They hunt you in three-dimensional space just as you hunt them. That doesn't happen in the movies, does it? Here you're the central actor, you're the star!"

Our answer combines hands-on demonstration with a reminder of other experiences. We draw a contrast, point- ing to something that VR is not. We still have not said what it is.

To answer what VR is, we need concepts, not samples or dictionary phrases or negative definitions.

OK, so what is it?

Our next reply must be more informed: "Go to the source. Find the originators of this technology; ask them. For twenty years, scientists and engineers have been working on this thing called *virtual reality*. Find out exactly what they have been trying to produce."

When we look to the pioneers, we see virtual reality going off in several directions. The pioneers present us with at least seven divergent concepts currently guiding VR re- search. The different views have built camps that fervently disagree as to what constitutes virtual reality.

Here is a summary of the seven:

Simulation

Computer graphics today have such a high degree of realism that the sharp images evoke the term *virtual reality*. Just as sound systems were once praised for their high fidelity, present-day imaging systems now deliver virtual reality. The images have a shaded texture and light radiosity that pull the eye into the flat plane with the power of a detailed etching. Landscapes produced on the GE Aerospace "visionics" equipment, for instance, are photorealistic real-time texture-mapped worlds through which users can navigate. These dataworlds spring from military flight simulators. Now they are being applied to medicine, entertainment, and education and training.

The realism of simulations applies to sound as well. Three-dimensional sound systems control every point of digital acoustic space, their precision exceeding earlier sound systems to such a degree that three-dimensional audio contributes to virtual reality.

Interaction

Some people consider virtual reality any electronic representation with which they can interact. Cleaning up our computer desktop, we see a graphic of a trash can on the computer screen, and we use a mouse to drag a junk file down to the trash can to dump it. The desk is not a real desk, but we treat it as though it were, virtually, a desk. The trash can is an icon for a deletion program, but we use it as a virtual trash can. And the files of bits and bytes we dump are not real (paper) files, but function virtually as files. These are virtual realities. What makes the trash can and the desk different from cartoons or photos on TV is that we can interact with them as we do with metal trash cans and wooden desktops. The virtual trash can does not have to fool the eye in order to be virtual. Illusion is not the issue. Rather, the issue is how we interact with the trash can as we go about our work. The trash can is real in the context of our absorption in the work, yet outside the computer work space we would not speak of the trash can except as a virtual trash can. The reality of the trash can comes from its handy place in the world woven by our engagement with a project. It exists through our interaction.

Defined broadly, virtual reality sometimes stretches over many aspects of electronic life. Beyond computer-generated desktops, it includes the virtual persons we know through telephone or computer networks. It includes the entertainer or politician who appears on television to interact on the phone with callers. It includes virtual universities where students attend classes on line, visit virtual classrooms, and socialize in virtual cafeterias.

Artificiality

As long as we are casting our net so wide, why not make it cover everything artificial? On first hearing the term *virtual reality*, many people respond immediately: "Oh, sure, I live there all the time." By this they mean that their world is largely a human construct. Our environment is thoroughly geared, paved, and wired—not quite solid and real. Planet Earth has become an artifice, a product of natural and human forces combined. Nature itself, the sky with its ozone layer, no longer escapes human influence. And our public life has everywhere been computerized. Computer analysis of purchasing habits tells supermarkets how high and where to shelve the Cheerios. Advertisers boast of "genuine simulated walnut."

But once we extend the term *virtual reality* to cover everything artificial, we lose the force of the phrase. When a word means everything, it means nothing. Even the term *real* needs an opposite.

Immersion

Many people in the VR industry prefer to focus on a specific hardware and software configuration. This is the model set for virtual reality by Sutherland, Fisher, Furness, and Brooks, before whom the term *virtual reality* did not exist, since no hardware or software claimed that name.

The specific hardware first called VR combines two small three-dimensional stereoscopic optical displays, or "eyephones"; a Polhemus head-tracking device to monitor head movement; and a dataglove or hand-held device to add feedback so the user can manipulate objects perceived in the artificial environment. Audio with three-dimensional acoustics can support the illusion of being submerged in a virtual world. That is, the illusion is immersion.

According to this view, virtual reality means sensory immersion in a virtual environment. Such systems, known primarily by their head-mounted displays (HMD) and gloves, were first popularized by Jaron Lanier's VPL (Virtual Programming Language) Incorporated. The HMD cuts off visual and audio sensations from the surrounding world and replaces them with computer-generated sensations. The body moves through artificial space using feedback gloves, foot treadmills, bicycle grips, or joysticks.

A prime example of immersion comes from the U.S. Air Force, which first developed some of this hardware for flight simulation. The computer generates much of the same sensory input that a jet pilot would experience in an actual cockpit. The pilot responds to the sensations by, for instance, turning a control knob, which in turn feeds into the computer, which again adjusts the sensations. In this way, a pilot can get practice or training without leaving the ground. To date, commercial pilots can upgrade their licenses on certain levels by putting in a certain number of hours on a flight simulator.

Computer feedback may do more than readjust the user's sensations to give a pseudoexperience of flying. The feedback may also connect to an actual aircraft, so that when the pilot turns a knob, a real aircraft motor turns over or a real weapon fires. The pilot in this case feels immersed and fully present in a virtual world, which in turn connects to the real world.

When you are flying low in an F-16 Falcon at supersonic speeds over a mountainous terrain, the less you see of the real world, the more control you can have over your aircraft. A virtual cockpit filters the real scene and represents a more readable world. In this sense, VR can preserve the human significance of an overwhelming rush of split-second data. The heads-up display in the cockpit sometimes permits the pilot to view the real landscape behind the virtual images. In such cases, the simulation is an augmented rather than a virtual reality.

The offshoots of this technology, such as the Waldern

The Essence of VR

arcade game, should not distract us—say the immersion pioneers—from the applications being used in molecular biology (docking molecules by sight and touch), airflow simulation, medical training, architecture, and industrial design. Boeing Aircraft plans to project a flight controller into virtual space, so that the controller floats thousands of feet above the airport, looking with an unobstructed view in any direction (while actually seated in a datasuit on the earth and fed real-time visual data from satellite and multiple camera viewpoints).

A leading model of this research has been the workstation developed at NASA–Ames, the Virtual Interface Environment Workstation (VIEW). NASA uses the VIEW system for telerobotic tasks, so that an operator on earth feels immersed in a remote but virtual environment and can then see and manipulate objects on the moon or Mars through feedback from a robot. Immersion research concentrates on a specific hardware and software configuration. The immersive tools for pilots, flight controllers, and space explorers are a much more concrete meaning of VR than is the vague generalization "everything artificial."

Telepresence

Robotic presence adds another aspect to virtual reality. To be present somewhere yet present there remotely is to be there virtually (!). Virtual reality shades into telepresence when you are present from a distant location—"present" in the sense that you are aware of what's going on, effective, and able to accomplish tasks by observing, reaching, grabbing, and moving objects with your own hands as though they were close up. Defining VR by telepresence nicely excludes the imaginary worlds of art, mathematics, and entertainment. Robotic telepresence brings real-time human effectiveness to a real-world location without there being a human in the flesh at that location. Mike McGreevy and Lew Hitchner walk on Mars, but in the flesh they sit in a control room at NASA–Ames.

Telepresence medicine places doctors inside the patient's body without major incisions. Medical doctors like Colonel

Richard Satava and Dr. Joseph Rosen routinely use telepresence surgery to remove gall bladders without the traditional scalpel incisions. The patient heals from surgery in one-tenth the usual time because telepresence surgery leaves the body nearly intact. Only two tiny incisions are needed to introduce the laparoscopic tools. Telepresence allows surgeons to perform specialist operations at distant sites where no specialist is physically present.

By allowing the surgeon to be there without being there, telepresence is a double-edged sword, so to speak. By permitting immersion, telepresence offers the operator great control over remote processes. But at the same time, a psycho-technological gap opens up between doctor and patient. Surgeons complain of losing hands-on contact as the patient evaporates into a phantom of bits and bytes.

Full-Body Immersion

About the same time that head-mounted displays appeared, a radically different approach to VR was emerging. In the late 1960s, Myron Krueger, often called "the father of virtual reality," began creating interactive environments in which the user moves without encumbering gear. Krueger's is come-as-you-are VR. Krueger's work uses cameras and monitors to project a user's body so it can interact with graphic images, allowing hands to manipulate graphic objects on a screen, whether text or pictures. The interaction of computer and human takes place without covering the body. The burden of input rests with the computer, and the body's free movements become text for the computer to read. Cameras follow the user's body, and computers synthesize the user's movements with the artificial environment.

I see a floating ball projected on a screen. My computer-projected hand reaches out and grabs the ball. The computer constantly updates the interaction of my body and the synthetic world that I see, hear, and touch.

In Krueger's Videoplace, people in separate rooms relate interactively by mutual body painting, free-fall gymnastics, and tickling. Krueger's Glowflow, a light-and-sound room, responds to people's movements by lighting phosphorescent tubes and issuing synthetic sounds. Another environment, Psychic Space, allows participants to explore an interactive maze in which each footstep corresponds to a musical tone, all produced with live video images that can be moved, scaled, and rotated without regard to the usual laws of cause and effect.

Networked Communications

Pioneers like Jaron Lanier accept the immersion model of virtual reality but add equal emphasis to another aspect that they see as essential. Because computers make networks, VR seems a natural candidate for a new communications medium. The RB2 (Reality Built for Two) System from VPL highlights the connectivity of virtual worlds. In this view, a virtual world is as much a shared construct as a telephone is. Virtual worlds, then, can evoke unprecedented ways of sharing, what Lanier calls "post-symbolic communication." Because users can stipulate and shape objects and activities of a virtual world, they can share imaginary things and events without using words or real-world references.

Accordingly, communication can go beyond verbal or body language to take on magical, alchemical properties. A virtual-world maker might conjure up hitherto unheard-of mixtures of sight, sound, and motion. Consciously constructed outside the grammar and syntax of language, these semaphores defy the traditional logic of verbal and visual information. VR can convey meaning kinetically and even kinesthetically. Such communication will probably require elaborate protocols as well as lengthy time periods for digesting what has been communicated. Xenolinguists will have a laboratory for experiment when they seek to relate to those whose feelings and world views differ vastly from their own.

■ ■ ■

All right, enough!" shouts our questioner, bleary eyed with information overload.

"I've taken your virtual-reality tour, listened to the pioneers, and now my head is spinning. These pioneers do indeed explore in different directions. There's a general drift here but no single destination. Should I go home feeling that the real virtual reality does not exist?"

Let's not lose stamina now. We cannot let the question fizzle. Too much depends on searching for the true virtual reality.

We should not get discouraged because a mention of reality, virtual or otherwise, opens several pathways in the clearing.

Let us recall for a moment just how controversial past attempts were to define the term *reality*. Recall how many wars were fought over it.

People today shy away from the R-word. *Reality* used to be the key to a person's philosophy. As a disputed term, *reality* fails to engage scientific minds because they are wary of any speculation that distracts them from their specialized work. But a skeptical attitude will fall short of the vision and direction we need.

Here's a brief sidebar on how controversial the R-word has been throughout Western history:

■ Plato holds out ideal forms as the "really real" while he denigrates the raw physical forces studied by his Greek predecessors. Aristotle soon demotes Plato's ideas to a secondary reality, to the flimsy shapes we abstract from the really real— which, for Aristotle, are the individual substances we touch and feel around us. In the medieval period, real things are those that shimmer with symbolic significance. The biblical–religious symbols add superreal messages to realities, giving them permanence and meaning, while the merely material aspects of things are less real, merely terrestrial, defective rubbish. In the Renaissance, things counted as real that could be counted and observed repeatedly by the senses. The human mind infers a solid material substrate underlying sense data but the substrate proves less real because it is less quantifiable and observable. Finally, the modern period attributed reality to atomic matter that has internal dynamics or energy, but soon the reality question was doomed by the analytical drive of the sciences toward complexity and by the plurality of artistic styles.

This reminder of metaphysics should fortify us for the long haul. If for two thousand years Western culture has puzzled over the meaning of reality, we cannot expect ourselves in two minutes, or even two decades, to arrive at the meaning of virtual reality.

The reality question has always been a question about direction, about focus, about what we should acknowledge and be concerned with. We should not therefore be surprised when VR proves controversial and elusive. Creating a new layer of reality demands our best shot, all our curiosity and imagination, especially since for us, technology and reality are beginning to merge.

When we look for the essence of a technology, we are engaging in speculation, but not in airy speculation. Our speculation involves where we plant our feet, who we are, and what we choose to be. Behind the development of every major technology lies a vision. The vision gives impetus to developers in the field even though the vision may not be clear, detailed, or even practical. The vision captures the essence of the technology and calls forth the cultural energy needed to propel it forward. Often a technological vision taps mythic consciousness and the religious side of the human spirit.

Consider for a moment the development of space technology. (Keep in mind that an inner connection exists between outer space and cyberspace, as I will point out later.) The U.S. space program enjoyed its most rapid development in the 1960s, culminating in the moon walk in 1969. What was the vision behind it?

The U.S. space program was a child of the cold war. The May 1961 speech by President John F. Kennedy that set NASA's goals incorporated traditional elements of myth: heroic struggle, personal sacrifice, and the quest for national prominence. Yet the impetus for Kennedy's speech came largely from without. What launched the U.S. space program was the fear of being surpassed by the Soviets, who had made a series of bold advances in human space travel. The goal of the moon landing was for the United States an attempt not to be overtaken by the Soviet developments in manned space exploration.

Few Americans know about the vision of their Russian competitors in space exploration. Everyone knows, of course, that the Communist revolution in 1917 froze Russian public goals in the hackneyed single-party language of a Marxist-Leninist agenda. Some historians know the name of the great Russian rocket pioneer Konstantin Tsiolkovsky (1857–1935), who stands with the American Robert H. Goddard (1882–1945) and the German-born Hermann Oberth (b. 1894). But less is known about the background of Tsiolkovsky's thinking and the visionary philosophy that influenced the first generation of Soviet space explorers.

What lay behind the energetic push to send human beings into outer space? The Russians to this day have gathered far more data on human survival in outer space. The need for information was more than curiosity or a vague lust for new frontiers; it was a moral mission, a complex and imaginative grasp of human destiny in the cosmos. The early Russian rocket pioneers, who gave the impetus to the program, felt there was an essence to their space technology, a deep inner fire that inspired and directed the research. They felt an existential imperative that drew on the religious and cultural traditions coming down through the main stream of Russian history. This essence was not itself technological, and so we might call it the esoteric essence of space technology, the hidden core of ideas that in themselves are not technological. In fact, the ideas behind the first space exploration were lofty, awe inspiring, and even mystical.

The visionary ideas fueling Tsiolkovsky and the early Russian explorers came from N. F. Fedorov. Nikolai Fedorovich Fedorov (1828–1903) was a powerful inspiration to Soloviev, Dostoevsky, Tolstoy, and a whole generation of Russians who sought to understand how modernization connects with traditional religion and culture. Even the engineers of the Trans-Siberian Railway came often to sit at the feet of the famous sage. Fedorov lived an intensely spiritual life, dedicated exclusively to ideas and learning. His profound vision applied certain strands of Russian Orthodox spirituality to the harnessing of modern technology.

Sketching a national vision, Fedorov drew large. He argued that Russia should marshal its military and national strength toward a single goal: the conquest of nature. Conquering nature meant regulating the earth as a harmonious system. It meant controlling the weather so that harvests would be plentiful. It meant balancing nature so that all life-forms could thrive together in harmony. In his vision, Fedorov saw armies producing solar energy and harnessing the electromagnetic energy of the earth, using the energy to regulate the earth's motion in space, turning the earth into a vessel for cosmic cruises. Overpopulation would cease to be a problem as humanity colonized other planets.

Unique to Fedorov's vision is its guiding moral spark. Instead of basing the conquest of nature on dominance, aggression, and egoism, Fedorov shunned the notion that humans should rule the cosmos out of a selfish desire for material wealth and abundance. Instead, he envisioned the conquest of nature as an act of altruism. But being generous to future generations can be less than purely altruistic, for they can return the favor by their acclaim of our deeds. We must regulate the forces of nature, he believed, so altruistically that we serve those who cannot possibly return our favors: we must conquer nature in order to resurrect our ancestors, the ultimate act of altruism.

The resurrection of all our dead ancestors, and it alone, provides a lofty enough ideal to mobilize humanity to explore the entire universe, including outer space. Fedorov found this thought in Russian Orthodox Christianity. According to Christian belief, the dead will rise again so that Christ, in a final judgment, will reorganize and completely redeem the world. The bodies of all human beings will one day rise again, and this resurrection, according to Fedorov, will take place through the work of human beings who carry out the divine plan. The long-range goal of human cooperation must be to discover the laws of nature to such a depth that we can eventually reconstitute the bodies of past human beings from their remaining physical particles still floating about in the universe.

Fedorov's strategy was to channel science and technology toward the reunion of all humanity. He decried the heartless positivism that builds on the sufferings and corpses of previous generations, instead seeking a purely idealistic motive. Without such a high aim, a heartless science would ultimately turn against society. For him, and for the many Soviet scientists inspired by him, the ultimate aim of the space program was, quite literally, nothing less than resurrecting the dead.

Contrast this sublime—and to us incredible and bizarre—vision of the space program with current U.S. public policy. "The commercialization of space," as promoted by administrations since the late 1970s, offers civilian entrepreneurs new opportunities for investment. To cover this naked self-interest, a mythic notion from U.S. history adds the sense of a new frontier. As a mere resource for commerce, space holds little allure, but a new frontier beyond earth adds adventure to the hope for personal gain. The vision even draws on the California gold rush in the nineteenth century, the spirit of enterprise.

In fact, this last word, *enterprise*, shows us where the commercialization of space falls short. Commercialization fails to touch the essence of space exploration, for commercial interests will neglect the long-range research needed for space science. Commercialization also drives up the cost of information derived from space exploration so that the data from space will not be available to small businesses, university scientists, farmers, state and local governments, and developing countries. In short, this kind of exploration envisions no future, only short-range profit.

But for NASA, for space enthusiasts, and for the Pentagon people, *enterprise* has a capital *E*. The word refers to a spirit of business adventure, but it also, in many minds, has another important meaning. Many technical people today also take *enterprise* to be the proper name in a science fiction myth, that of the starship *Enterprise* in "Star Trek," the popular science fiction television series about twenty-first-century space travelers. "Star Trek" contributed the code word, the handshake, the common inspiration for space exploration in the United States. (Shake hands informally with someone at the Pentagon or NASA and be prepared with an answer to the query "Are you a Trekkie?") For hundreds of technicians, the

space program flies on the imaginative wings of Gene Roddenberry's brainchild, born on September 8, 1966, when the TV
show was first aired. But Roddenberry was no Fedorov. The
sage of Pasadena created no unifying vision to direct humanity
"where no one has gone before." His fictional productions
treated only a motley collection of profound moral questions
pertaining to human behavior at any time, any place. But despite the limits of its lineage, "Star Trek" showed us more
truly the esoteric essence, the real meaning, of space exploration than did government statements on the commercialization
of space. The essence of the American space program, its heart
and soul, comes from "Star Trek."

Where in VR is a counterpart to the space program's esoteric essence? What is the essence of VR, its inner spirit, the
cultural motor that propels the technology? When the first
conferences met on cyberspace and on virtual reality in 1989
and 1990, respectively, two threads of shared vision ran
through the diverse groups of participants. One was the cyberpunk writings of William Gibson, known to both technical
and literary types as the coiner of the term cyberspace. The
other was the Holodeck from "Star Trek: The Next Generation."

Along with its cargo bay of imaginative treasures, the
starship Enterprise brought the Holodeck. The Holodeck is
familiar furniture in the vocabulary of virtual-reality pioneers.
For most people, the Holodeck portrays the ideal human–
computer interface. It is a virtual room that transforms spoken
commands into realistic landscapes populated with walking,
talking humanoids and detailed artifacts appearing so lifelike
that they are indistinguishable from reality. The Holodeck
is used by the crew of the starship Enterprise to visit faraway
times and places such as medieval England and 1920s
America. Generally, the Holodeck offers the crew rest and
recreation, escape and entertainment, on long interstellar voy-
ages.

While not every VR pioneer explicitly agrees on goals,
the Holodeck draws the research onward. Publicly, researchers
try to maintain cool and reasonable expectations about VR.
Hyperbole from the media often stirs grandiose expectations
in the public; when presented with actual prototypes, the pub-

lic turns away with scorn. So researchers play down talk of the Holodeck. At the MIT Media Lab, leaders such as David Zeltzer avoid the term *virtual reality* not only because of the specter of metaphysics it evokes, but also because of the large promises it raises. The term seems to make greater claims than do terms like *virtual environments* (preferred at MIT and NASA) and *virtual worlds* (preferred at the universities of North Carolina and Washington). But when speaking at a VR conference for the Data Processing Management Association in Washington, D.C., on June 1, 1992, Zeltzer made an intriguing aside, one that touches, I think, on the highest possibilities of virtual reality, on its esoteric essence.

Did I say "esoteric essence"? How can we expect to give our young questioner an answer to "What is virtual reality?" when we have left the public, exoteric world of clear explanations and have embarked on a search for the esoteric essence of VR, its underlying vision? Well, our questioner seems to have gotten lost some time ago, most likely during the sidebar on the history of reality. I think I see someone off in the distance pulling avidly on the trigger of the Virtuality game. Maybe more time spent in VR will eventually deliver better answers than any verbal speculation. At any rate, on to the esoteric essence . . .

Zeltzer's remark went something like this: "True virtual reality may not be attainable with any technology we create. The Holodeck may forever remain fiction. Nonetheless, virtual reality serves as the Holy Grail of the research."

"Holy Grail?" Holy Grail!

Now when Zeltzer made this reference, he was not deliberately invoking a Jungian archetype. His remark expressed modesty and diffidence rather than alchemical arrogance. Still, archetypes do not have to hit us in the nose to wield their peculiar power. They work most powerfully at the back of the subconscious mind, and therein lies their magic. An effective archetype works its magic subtly.

David Zeltzer was calling up a mythic image far more ancient and infinitely more profound than "Star Trek." "Star Trek" has, after all, become the stuff of trivia: "Star Trek" ties and boxer shorts, "Star Trek" vinyl characters and mugs

("Fill them with a hot beverage and watch Kirk and Spock beam up to an unknown world"). "Star Trek" lost any sublimity it may have had when it came to occupy Kmart shelves along with electric flyswatters and noisemaker whoopee cushions.

The Holy Grail, though, sums up the aspirations of centuries. It is an image of the Quest. From Tennyson's romantic *Idylls of the King* to Malory's King Arthur and the Knights of the Round Table, the ancient Grail legend reaches back to Christian and pre-Christian times. The Grail has always been a symbol of the quest for a better world. In pre-Christian times, the Grail was the cup that holds a cure for an ailing king who, suffering from his own wounds, sees his country turning into a wasteland. Christians believed the Grail to be both the chalice of Jesus' Last Supper and the cup that caught the Savior's blood at the Crucifixion. Medieval legend links the spear that pierced Jesus' side on the cross with the sacred cup that held his blood. Later works of art, from T. S. Eliot's *The Wasteland* to Richard Wagner's *Parsifal*, have preserved the Grail story as a symbol of spiritual quest and lofty aspiration.

Perhaps the essence of VR ultimately lies not in technology but in art, perhaps art of the highest order. Rather than control or escape or entertain or communicate, the ultimate promise of VR may be to transform, to redeem our awareness of reality—something that the highest art has attempted to do and something hinted at in the very label *virtual reality*, a label that has stuck, despite all objections, and that sums up a century of technological innovation. VR promises not a better vacuum cleaner or a more engrossing communications medium or even a friendlier computer interface. It promises the Holy Grail.

We might learn something about the esoteric essence of VR by thinking about Richard Wagner's *Parsifal*. Wagner himself was searching for a Holodeck, though he did not know it. By the time he finished *Parsifal*, his final opera, Wagner no longer considered his work to be opera. He did not want it called opera or music or theater or even "art," and certainly not entertainment. By the time he finished his last work, Wagner realized he was trying to create another reality, one that would in turn transform ordinary reality. The term he came to use was "a total work of art," by which he meant a seamless union of vision, sound, movement, and drama that would sweep the viewer to another world, not to escape but to be changed. Nor could the viewer be a mere spectator. Wagner created a specially designed building in Bayreuth, Germany, well off the beaten track, where the audience would have to assemble after a long journey because he forbade the performance of *Parsifal* in any other building. The audience would have to prepare itself well ahead of time by studying the libretto, because *Parsifal* was long, mysterious, and full of complex, significant details. (Wagner's *Ring* cycle takes over fifteen hours to present a related myth.) Looking for the right terms to express his intent, Wagner called *Parsifal* "a festival play for consecrating the stage" (*ein Bühnenweihfestspiel*). The Bayreuth theater would become the site for a solemn, nearly liturgical celebration. The mythmaker would create a counter-reality, one reminiscent of the solemn mass of the Catholic church, which appeals to all the senses with its sights, sounds, touch, drama, even appealing to smell with incense and candles. The audiences at Bayreuth were to become pilgrims on a quest, immersed in an artificial reality.

The drama *Parsifal*, like a mysterious dream, resists easy summary, and it eludes interpretation. But the general story outline is clear. The protectors of "correct values" (the Knights) inevitably paint themselves into the corner of righteousness. Paralyzed, unable to act, their leadership suffers intense internal pain (Amfortas). They can regain the power of the Grail that they protect only through the intervention of someone who is still innocent of right and wrong, someone who is by all standards a fool. The innocent fool (Arabic, *fal parsi*) can clean out the sclerotic righteous society only after passing a test and learning to feel the sufferings of others. Once the innocent fool has acquired compassion for others and sensitivity to life's complexity, he can bring the power (the Spear) back to the righteous Knights of the Holy Grail. The Grail Knights then come to understand more deeply what the work of the Holy Grail, and their mission, means. The

Grail grants its full power only to those who can be touched by compassion.

Wagner's Holodeck presents a Parsifal who mirrors the individual audience members at Bayreuth. Wagner shaped the drama with story and music so that strong sensations would engulf the audience and pierce them to the heart. Each listener begins as a naive spectator and is then gradually touched by the painful actions on stage until the listener becomes transformed into a more sensitive and compassionate member, ready to bring to a sick society some measure of healing and renewal.

Wagner hoped to do more than make music and theater; he believed that his music dramas could transform society by imparting new feelings and attitudes. This goal he shared with traditional religion; and religion returns the competition with distrust and the accusation of heterodoxy. For this reason, Wagner's work remains to this day controversial among religious people, including many artists and musicians who have strong religious faith.

How well did Wagner succeed? One of the most telling tributes to the success of Wagner's *Parsifal* comes from a Jesuit priest, Father Owen Lee, who in a radio broadcast intermission feature from the Metropolitan Opera in New York City said:

> I watched as usual from the least expensive seat under the roof, hovering there with an unearthly feeling for long half-hours floating in an immense space, suffused with a sense of what Baudelaire felt listening to Wagner: "A sense of being suspended in an ecstasy compounded of joy and insight." I can remember staggering out of theaters after *Parsifal*, hardly aware of people applauding, the music streaming through me, carried out of myself, seeing my experience—indeed, feeling that I was seeing all experience—at a higher level of awareness, put in touch with a power greater than myself, a kind of holy fool.[1]

Another holy fool was the Finnish composer Jan Sibelius, who wrote: "Heard *Parsifal*. Nothing else in all the world has made so overwhelming an impression on me. All my heartstrings throbbed." The German composer Max Reger wrote: "Heard

Parsifal. Cried for two weeks, then decided to become a composer."

Someday VR will elicit similar rave reviews, not mere thrills, but insight into experience. As it evolves its art form, VR will have certain advantages over Wagner's "total work of art." Certain disadvantages might also plague it where Wagnerian solutions might help.

Activity/Passivity

VR systems, as Jaron Lanier points out, can reduce apathy and the couch-potato syndrome simply by requiring creative decisions. Because computers make VR systems interactive, they also allow the artist to call forth greater participation from users. Whereas traditional art forms struggle with the passivity of the spectator, the VR artist finds a controlled balance between passivity and activity. The model of user navigation can be balanced by the model of pilgrimage and sacred awe.

Manipulation/Receptivity

Some observers date the advent of VR to the moment when the dataglove appeared on the computer screen. At that moment, the user became visible as an active, involved force in the digital world. This implies that VR has a tilt toward manipulation, even a latent tendency toward aggressive, first-person attitudes. The VR artist will need strategies for inducing a more receptive atmosphere, so that the user can be open in all directions, receiving signals from and having empathy for other beings. The user must be able to be touched, emotionally moved, by non-first-person entities in the virtual world. The spear of manipulation must join the cup of sensitivity. If simulators serve to train hand–eye and other coordination skills, VR may take a further step and become a training tool to enhance receptivity.

Remote Presence

The visual bias of current VR brings out a possible detachment in the user's sense of the world. Seeing takes place at a distance, whereas hearing and the other senses are more intimate to our organic life. The visual bias increases the detachment of telepresence. Some VR versions stress the "looking-at" factor, such as David Gelernter's Mirror Worlds, in which, in real-time, users can zoom in on miniature shoe-box worlds containing local homes, businesses, cities, governments, or nations. VR offers the opportunity to shift the Western philosophy of presence. From Pythagoras to Aristotle, from Berkeley to Russell, our philosophical sense of presence has relied on vision, consequently putting us in the position of spectators. To be touched, we need to introduce more sensory awareness. VR may develop a kind of feedback in which presence includes an openness and sensitivity of the whole body.

Augmented Reality

VR will enhance the power of art to transform reality. The picture frame, the proscenium, the movie theater all limit art by blocking it off as a section of reality. VR, with its augmented reality, allows a smoother, more controlled transition from virtual to real and back. This capability, which may frighten psychologists, will offer artists an unprecedented power to transform societies.

■ ■ ■

These are a few of the differences that make virtual reality different from traditional art forms. They belong to the essence of VR, its Holy Grail. This goal means that we need a different breed of artist as well. And where will we find these new cybersages, these virtual-world makers? I see our young questioner smiling broadly now as yet another wounded pterodactyl drops from the pink sky of Waldern's arcade game. Plenty of fledgling enthusiasm here, and a society that needs healing and renewal.

Note

1. Father Owen Lee, "Metropolitan Opera Broadcast Intermission Feature," March 28, 1992.

9

VIRTUAL-REALITY CHECK

The commission money was good, and the artist arrived on time. One of the executives from corporate design was there to meet her at the door. After touring the facilities, the artist was left alone to begin painting. Each day the mural materialized a bit more, section by section, spreading a ribbon of color across the large gray wall at the end of the lobby. First a green patch of forest glade appeared, two blossoming plum trees, three sky blue vistas, and a Cheshire cat on a branch. Finally came the day when the tarp would fall. Employees gathered around plastic cups and croissants. When the speeches were over, the room grew hushed for the unveiling. The crowd gasped. The wall came alive with paradise, an intricate world of multicolored shapes. Several employees lingered to chat with the artist. Once the congratulations died down, the artist strolled to the center of the mural, stopping where the garden path leads into the forest, and with her face to the crowd, she smiled, bowed, and turned her back. Walking into the green leaves, she was never seen again.

This ancient story, adapted from Taoist legend, anticipates the metaphysics of virtual reality. On one level, the story praises the power of artistic illusion. On a deeper level, it suggests our need to create realities within realities, to suspend our belief in one set of involvements in order to entertain another. The story depicts our ability to enter symbolic space, where we move about in alternate worlds. Whether we read a short story, watch a film, or contemplate a painting, we enjoy being hijacked to another plane of being. Our capacity

to immerse ourselves in a symbolic element has developed
to the point that we hardly even notice the disappearing act.
We slip off into symbolic existence at the drop of "Once upon
a time" or "Given any variable X."

From Naive Realism to Irrealism

Are not all worlds symbolic? Including the one we naively
refer to as the real world, which we read off with our physical
senses? Philosophers as recent as Nelson Goodman and Richard Rorty have considered all worlds—not just the world of
storytelling and filmmaking—to be contingent symbolic constructs. Science, religion, and art provide different versions
that are made, tested, and known in diverse ways, each with
its own rightness and function. Each world is made from the
previous world(s), and each process of world making proceeds
by composing or decomposing older materials, by identifying
repetitions and evolving new patterns, by deleting and supple-
menting, and by organizing and ordering aspects of the
world(s) already there. One well-worn way to point out this
diversity is to compare the weather vocabulary of the Eskimos
with that of the surfers in southern California.

When did the universe break into a plurality? Since Immanuel Kant, philosophy has moved gradually from the
unique reality of a single fixed world to a diversity of worlds.
Kant eliminated the notion of a pregiven world by locating
orderly patterns not in the found world, but in the architecture
of the human mind. The categories of the understanding (causality and substance) along with the forms of intuition (space
and time) mold the chaotic givens of perception, forging an intelligible, communicable structure of experience. Still, Kant
postulated a monistic ideal of unity to regulate our construction of the world. The world we make, he believed, drives
toward a single shared unity. In this way, Kant protected the
Newtonian science of his era by basing knowledge on the supposedly absolute forms lodged in human reasoning. After
Kant, philosophers whittled away at the monistic unity until
quantum theory in the twentieth century withdrew support
for the kind of coherence Kant thought essential to science.

Now, with science itself open to diversity and indeterminateness, many philosophers welcome the world as a plurality. In our day, Nelson Goodman, for instance, claims: "Our passion for *one* world is satisfied, at different times and for different purposes, in *many* different ways. Not only motion, derivation, weighting, order, but even reality is relative."[1] Goodman's *Ways of Worldmaking*, in which he promotes the doctrine of irrealism, seems a proper primer for the architects of virtual reality.

Realism and Irrealism: Both Unrealistic

Irrealism may be shortsighted. We may need to hang onto a notion of the real world, if not out of abstract conviction then at least out of the need for occasional reality checks against our virtual-reality systems. Virtual worlds can threaten the integrity of human experience. We see how technologies disrupt our biobodies in the examples of jet lag and flight simulator sickness. The cyberbodies of virtual reality may further upset an already precarious ontological balance. The modern person's typical body amnesia may deepen as Alternate World Syndrome and Alternate World Disorder begin to appear (see the glossary under AWS and AWD.) We need to learn how to do occasional virtual reality checks. An unrestrained proliferation of worlds cries out for sanity, for connection with reality, for metaphysical grounding.

Kant dismissed metaphysical theories as idle sophistries and intellectual games played by charlatans. Philosophers in the twentieth century, from Wittgenstein and Heidegger to Carnap and Ayer, followed Kant in sidestepping metaphysics, believing it to be an ungrounded spin of the linguistic wheels, a chase after vague vapor trails, or simply a logical mistake. For this line of thinking, *reality* has lost its meaning as a serious term. The coming VR engines may force a change in that general line of thought and shed new light on classical metaphysics. The next century may have to dig again in a very ancient field of metaphysics excavated by the engines of computer-simulated virtual reality, the metaphysical machine par excellence. Conversely, virtual realities may be all the richer for preserving some relationship to a real world, without, however, becoming boring or mundane. The terms *real* and *virtual* need sorting out before we relate them to each other. I find the contemporary usage of the word *virtual*, as well as that of its distant ancestor, enlightening.

The Vocabulary of Virtuality

In contemporary usage, the *virtual* in *virtual reality* comes from software engineering. Computer scientists use *virtual memory* to mean computer RAM set aside in such a way that the computer operates as though memory exists beyond the actual hardware limits. The term *virtual* has come to connote any sort of computer phenomenon, from virtual mail to virtual work groups on computer networks, to virtual libraries, and even to virtual universities. In each case, the adjective refers to a reality that is not a formal, bona fide reality. When we call cyberspace a virtual space, we mean a not-quite-actual space, something existing in contrast with the real hardware space but operating as though it were real space. Cyberspace seems to take place within the framework of real space.

The *virtual* in *virtual reality* goes back to a linguistic distinction formulated in medieval Europe. The medieval logician John Duns Scotus (1266?–1308) gave the term its traditional connotations. His Latin *virtualiter* served as the centerpiece of his theory of reality. The Doctor of Subtlety maintained that the concept of a thing contains empirical attributes not in a formally (as though the thing were knowable apart from empirical observations) but *virtualiter*, or virtually. Although we may have to dig into our experiences to unveil the qualities of a thing, Duns Scotus held, the real thing already contains its manifold empirical qualities in a single unity, but it contains them virtually—otherwise they would not stick as qualities of that thing. Duns Scotus used the term *virtual* to bridge the gap between formally unified reality (as defined by our conceptual expectations) and our messily diverse experiences. Similarly, we use the term *virtual* to breach the gap between a given environment and a further level of artificial accretions. Virtual space—as opposed to natural bodily space—contains the informational equivalent of things. Virtual space makes us feel as if we were dealing directly with physical or natural realities. As if . . .

Our "as if" stops short of Scotus's term, for he could assume, as could all classical metaphysicians, that our concepts fit squarely with the eternally fixed essences of things. Duns Scotus could assign a merely virtual reality to some aspects of experience because he believed that his primary experience already exhibited "real reality," to use Plato's strange phrase. Classical and medieval philosophy equated reality with the permanent features of experience, and this naive realism anchored human beings in the world. The medievals believed that the anchor held with all the weight of an all-powerful, unchanging God.

We cannot locate the anchor for our reality check outside this fluctuating, changing world. No universal divinity ensures an invariant stability for things. But we need some sense of metaphysical anchoring, I think, to enhance virtual worlds. A virtual world can be virtual only as long as we can contrast it with the real (anchored) world. Virtual worlds can then maintain an aura of imaginary reality, a multiplicity that is playful rather than maddening.

A virtual world needs to be not-quite-real or it will lessen the pull on imagination. Something-less-than-real evokes our power of imaging and visualization. Recall the legend of the vanishing artist. The magic in the story comes from the crossover of three-dimensional to two-dimensional frameworks. On another level, the magic of the story comes from our ability to cross over from the words of the narration to an inner vision of the sequence of virtual events (which occurs in us as we walk through the wall of words on the page). The story relates a legend about the power of symbols while exhibiting that power. Imagination allows us to use what we read or hear to reconstitute the symbolic components into a mental vision. The vision transcends the limits of our bodily reality, so that from the viewpoint of bodily existence, imagination is an escape, even though imagination often introduces new factors into our lives that sometimes cause us to alter our actual circumstances.

Virtual-Reality Check

For the most part, imagination receives in order to create. We use the words of a story or the photos of a film to reconstitute their contents, customizing the narrative details to our own understanding. Especially when using a single sense like hearing or touch, we are active in receiving information. All the other senses subconsciously join to reconstitute the content. But imagination always leaves behind the limits of our physical existence, and for this reason it is "only" imagination. Because it leaves the real world behind, imagination is not reality. When the artist takes her body with her through the mural painting, it is our imagination (through the story) that completes her work of art.

The Virtues of Cyberspace

Cyberspace, too, evokes our imagination. Cyberspace is the broad electronic net in which virtual realities are spun. Virtual reality is only one type of phenomenon in electronic space. As a general medium, cyberspace invites participation. In the framework of the everyday world, cyberspace is the set of orientation points by which we find our way around a bewildering amount of data. Working on a mainframe computer, like the Cyber 960 or the VAX 6320, we must learn to sketch a mental map for navigating the system. Without a subconsciously familiar map, we will soon lose our way in the information wilderness. Using a desktop or portable computer requires a similar internal depiction of how the hardware, CRT, keyboard, and disk drives connect, even if the picture is mythical or anthropomorphic, just as long as it works. Magnetic storage offers no three-dimensional cues for physical bodies, so we must develop our own internally imaged sense of the data topology. This inner map we make for ourselves, plus the layout of the software, is cyberspace.

The familiar mental map can be compared with the full-featured virtual reality as radio is with television or television with three-dimensional bodily experience. In its simplest form, cyberspace activates the user's creative imagination. As it becomes more elaborate, cyberspace develops real-world simulations and then virtual realities. William Gibson's cyberspace presents the data of the international business community as a three-dimensional video game. Gibson's users become involved through a computer console connected by electrodes feeding directly into the brain. The user's body stays behind to punch the deck and give the coordinates while the user's mind roams the computer matrix. The user feels the body to be "meat," or a chiefly passive material component of cyberspace, while the on-line mind lives blissfully on its own. *Neuromancer* describes the passivity of cyberspace as "a consensual hallucination. . . . A graphic representation of data abstracted from the banks of every computer in the human system. Unthinkable complexity." On the active side, the user pursues "the lines of light ranged in the nonspace of the mind."[2]

Virtual Realities
Without Ontological Security

The problem is not with cyberspace, but with virtual reality. I do not have to imagine myself bodily entering a virtual world. The computer's VR will soon allow me to take my body along with me, with either a sensorium interface or a third-person iconic representation. The degree of realism is, in principle, unlimited. This very realism may turn into irrealism, in which virtual worlds are indistinguishable from real worlds, virtual reality becomes bland and mundane, and users undergo predominantly passive experiences akin to drug-induced hallucinations. If Schopenhauer is right when he says that we are incorrigibly metaphysical animals, then this irrealism violates something we need and puts a possible limit on virtual-reality construction.

How may we preserve the contrast between virtual and real worlds? How can virtual realities preserve a built-in contrast with real or anchored reality so that we will enjoy a metaphysical pull to create and actively use our imaginations in cyberspace? What anchor can serve to keep virtual worlds virtual?

This is no place to launch a full-scale "battle of the Titans," as Plato described metaphysics. But I want to suggest some existential aspects of the real world that provide clues to preventing a virtual world from flattening out into a literal déjà vu. These existential features, evolving from twentieth-century philosophy, stand open to revision. Virtual worlds evoke imagination only if they do not simply reproduce the existential features of reality but transform them beyond immediate recognition. The existential features of the real world to which I refer include mortality/natality, carryover between past and future, and care.

The Three Hooks on the Reality Anchor

The real world, conceived existentially, functions with built-in constraints that provide parameters for human meaning. One constraint, our inevitable mortality, marks human existence as finite. Because of our limited life span, we demarcate our lives into periods of passage as well as into the schedules and deadlines that order our work flow. We are born at a definite time (natality) and grow up within distinct interactions (family kinships). These limits impose existential parameters on reality, providing us with a sense of rootedness in the earth (a finite planet with fragile ecosystems). Mortality/natality belongs to the reality anchor. Another reality constraint is temporality, the built-in carryover of events from the past into the future, our memory, or history. We can erase nothing in principle from what happens in a lifetime. What the German language calls *Einmaligkeit*, or "once-and-always-ness," endows actions with uniqueness and irretrievability. The carryover feature distinguishes reality from any passing entertainment or momentary hallucination. Finally, because of the temporariness of biological life-forms, a sense of fragility or precariousness pervades our real world, frequently making suffering a default value. The possibility of physical injury in the real world anchors us in an ultimate seriousness, the poignancy of which permeates casual phrases like "Take care." We care because we are fragile and have to be careful. These three features mark human existence and stamp experience with degrees of reality. They anchor us.

Should synthetic worlds, then, contain no death, no pain,

no fretful concerns? To banish finite constraints might disqualify virtuality from having any degree of reality whatsoever. Yet to incorporate constraints fully, as some fiction does, is to produce an empty mirror over and above the real world, a mere reflection of the world in which we are anchored. (I think of Bobby Newmark in Gibson's *Count Zero* and the dead boy who Wilson carried off Big Playground on a stretcher.) Actual cyberspace should do more; it should evoke the imagination, not repeat the world. Virtual reality could be a place for reflection, but the reflection should make philosophy, not redundancy. "Philosophy," said William James, "is the habit of always seeing an alternative."[3] Cyberspace can contain many alternate worlds, but the alternateness of an alternate world resides in its capacity to evoke in us alternative thoughts and alternative feelings.

Any world needs constraints and finite structure. But which aspects of the real (existential) world can attract our attention and sustain our imagination? Time must be built in, but the way of reckoning time need not duplicate the deadlines of the real world. Time could have the spaciousness of a totally focused project or could be reckoned by rituals of leisure. Danger and caution pervade the real (existential) world, but virtual reality can offer total safety, like the law of sanctuary in religious cultures. Care will always belong to human agents, but with the help of intelligent software agents, cares will weigh on us more lightly.

The ultimate VR is a philosophical experience, probably an experience of the sublime or awesome. The sublime, as Kant defined it, is the spine-tingling chill that comes from the realization of how small our finite perceptions are in the face of the infinity of possible, virtual worlds we may settle into and inhabit. The final point of a virtual world is to dissolve the constraints of the anchored world so that we can lift anchor—not to drift aimlessly without point, but to explore anchorage in ever-new places and, perhaps, find our way back to experience the most primitive and powerful alternative embedded in the question posed by Leibniz: Why is there anything at all rather than nothing?

Notes

1. Nelson Goodman, *Ways of Worldmaking* (Indianapolis: Hackett, 1978), p. 20.
2. William Gibson, *Neuromancer* (New York: Ace Books, 1984), p. 51.
3. William James, "Philosophy and the Philosopher," in *The Philosophy of William James*, drawn from his own works, with an introduction by Horace M. Kallen (New York: Modern Library, [1925]), p. 58.

10

THE ELECTRONIC CAFÉ
LECTURE

How pleasant and appropriate a spot for talking about virtual reality—Santa Monica's Electronic Café, where televised images reflect and transmit our living words and gestures.

The phone call inviting me to the first meeting of the Southern California Virtual Reality Special Interest Group brought to mind the movie *Ghostbusters*. I thought of Bill Murray and Dan Ackroyd in the movie, fighting the plasmic ghosts that descended on New York City. "Who ya gonna call?" went the gleeful Ghostbusters' song, mocking the confusion of a skeptical city come face-to-face with metaphysical monstrosities. Today, an antimetaphysical era has begun to fear dangerous spooks oozing from virtual-reality technology, and since the late 1980s, the philosophical phones have been ringing. Like Bill Murray, I don't know whether the job should make me laugh or cry. Not only do I see metaphysical questions behind VR technology, but more and more people, both inside and outside the computer industry, are seeing them too. These big questions might look silly and pale in the broad daylight of the present, but just dim the lights, glimpse the future, and . . . watch out! So, following the lead of the original Ghostbuster, I plug my tongue firmly in cheek, strap on the dissertation equipment, and march hesitantly forward.

When first thinking about computers, what most struck me was not the computer as a potential rival to our human intelligence. Human intelligence has been—and probably will always be—controversially hard to pinpoint, and computers neither notice patterns nor contemplate alternatives very well. The task for which computers are most effective is tracking linear sequences and storing the past as information; computers have no future, at least no biologically limited life span.

What first struck me about computers was how smoothly they integrate with human thought processes. Computer software can mirror the mind so closely that together human and computer constitute a third kind of entity, a marriage of person and machine, an augmented human intelligence.

As a writer, I first noticed the human–computer fusion when in 1983 I learned to use a word processor. I saw the impact of word processing on the way people write, read, and even think. My book *Electric Language* explored that transition between the world of hand- or typewritten manuscripts and the world of digital information. The change was so great that I called it an ontological shift, a change in our awareness of reality.

Academics generally ignored *Electric Language*, which was written in 1985, published in 1987, and became a paperback in 1989. Most academics were still installing software and figuring out how keyboard macros work. The reality shift I saw was not a highly visible surface break, but a deep, underlying, slow-moving drift in the underground tectonic plates of our awareness. The software industry—computer journals and information managers—took note of my speculations. They tended to agree that we are going through more than a change of tools. Some even testified to the same trade-offs that my book detailed in the changeover from the "psychic framework of the book" to the psychic framework of digital text. But my theory of the reality shifts underlying the computer revolution became more plausible with the computer breakthrough of the late 1980s: virtual reality.

By late 1989, the computer researchers were singing, "Who ya gonna call?" The metaphysical monsters had begun to appear. First, Jaron Lanier went public with VPL's eyephones and datagloves, coming on like the technoevangelist of virtual reality. Mike McGreevy and Lew Hitchner showed their virtual Mars walk at NASA–Ames. The Human Interface

Lab in Seattle took viewers over a simulated Port of Seattle. The video arcade games, especially those made by Jonathan Waldern, had grown to where they verged on the next step: full sensory immersion in a totally computer-generated environment. Fred Brooks, a twenty-year research leader in the field at the University of North Carolina, clearly showed that the next steps could be taken, although more time and work were needed to produce smooth computer-generated virtual environments.

The public had seen the Holodeck in "Star Trek: The Next Generation," and many people—including scientists, engineers, and futurologists—had read William Gibson's award-winning 1984 novel *Neuromancer*. Gibson had introduced the notion of cyberspace, a total electronic environment in which people can interact with data. The concept had arrived along with the elementary hardware and software. A new layer of reality seemed imminent, and it came announced in the very name of the technology: virtual reality.

"Who you gonna call?" The metaphysician's phone began ringing. The trade journal *Multimedia Review* wanted to instruct its readers about the philosophical problems that are puzzling the designers and users of VR. Michael Benedikt at the University of Texas put the ontological question of cyberspace (What is its reality connection?) at the top of the list when he convened the First Conference on Cyberspace in 1990. Soon the Education Foundation of the Data Processing Management Association and the Technology Training Corporation hired a metaphysician to organize conferences in Washington, D.C., to educate military, government, and commercial decision makers. Where was this new technology about to take us? How does virtual reality connect with real reality?

As the talk poured back and forth, I sifted out two main streams of interest in VR technology. The two streams split the United States down the middle, revealing a not-surprising bicoastal attitude gap stretching between the West and East Coasts of the United States. The West Coast warms up to an ecstatic, otherworldly, countercultural VR. In San Francisco, Howard Rheingold led the *Whole Earth Review* with a brilliant overview of actual VR experiments and his musings on virtual sex ("teledildonics"), which were snapped up by the glossy magazines *Esquire, M,* and *Playboy.* The East Coast convened Senate hearings and industry conferences to explore the military and industrial uses of VR. The National Security Council and the Department of Defense showed interest in the use of illusion and simulation. After all, most of the early groundwork for VR was done in the 1960s by the U.S. Air Force, designing flight simulators under Tom Furness at Wright–Patterson Air Force Base.

The two bicoastal attitudes toward artificial reality reflect a split in our basic human sense of reality. We want two things at once: we want to protect what we already possess, and at the same time we continually want to surpass what we already have. While we need established values, we also need to challenge the complacent satisfaction that accompanies our established systems. There are two coasts in the mind. The West Coast wants VR to serve as a machine-driven LSD that brings about a revolution in consciousness; the East Coast wants a new tool for supporting current projects and solving given problems.

Where is the truth about VR? Which coast sees the promise of VR more accurately? The truth, as usual, lies somewhere in the middle. The would-be counterculture correctly believes that VR will change us. Change us it will, but not in ways we can aim at directly or even conceive clearly. The changes will be not so much in what we want and desire, but in who we believe we are and what we think we are doing here. So the counterculture remains mired in past human wishes and desires. And establishment types who seek to adapt VR to already given ends are also correct. A splendid new tool is on the way, and it will indeed give us more power to protect ourselves and control our environment. But a tool like VR will teach us new tricks. We will have to adapt as it touches our deepest sense of what and where we are. The very notion of human presence is on the line here.

VR is bringing about something new in our relationship to technology in general. Certainly technology has been multiplying the ethical decisions to be made in areas like health care, with each specific techological advance raising new questions. But VR is the first technology to be born socially self-critical. Publicly debated at birth, VR is being talked about even though it is still in its early embryonic stages. Our information systems and electronic media, from MTV to USENET, are bringing VR to public attention even while the technology is part concept, part product. VR may be signaling a new relationship we have to technology in general. After all, is not VR the world reborn in artificial form? Previously, scientists did research and engineers created designs that then went into commercial production. Later the products went on to exert their effects on society at large. Now, even before seeing the products, we face the responsibility of projecting their impact and weighing the probable development of the technology. Many think this advance speculation is appropriate, for we are positioning ourselves to create whole worlds in which we will pass part of our lives.

Movies create places where we can sit together and watch common fantasies. They project communal anxieties so that we can later talk about them. Although they are neither interactive nor immersive, movies do prepare us to some extent for inhabiting virtual worlds by bringing us together to discuss possibilities. Some of the questions raised by VR that deserve public discussion are already coming through the projectors and loudspeakers at showings of the 1992 movie *Lawnmower Man*. I will conclude by pointing out for discussion some of the questions raised by Brett Leonard's movie. The questions that *Lawnmower Man* raises are many, perhaps too many, for discussion. I will limit my discussion to those themes I have dealt with previously in my writings, as in that way I can draw attention to the connection between the philosophical tradition and the emergence of VR.

Lawnmower Man begins with technoanxiety. Dr. Larry Angelo awakens from a nightmare, "a bad nightmare, really bad." His dream envisioned a runaway psychotic chimpanzee whose intelligence was amplified by guided virtual reality. VR offers the hope of human evolution. "Once the goal of magicians and alchemists," says Dr. Angelo, "we can now use virtual reality to actually bring the human being up to the next stage of evolution." Yet Caroline, Dr. Angelo's mate, regards the technology as a diversion from biologically grounded life, as a wasteful ignorance of natural needs: "Larry, this technology might be the future to you, but it's the same old shit to me." Larry Angelo feels his dream of human perfection being tossed back and forth between Frankenstein and Henry David Thoreau.

In Chapter 9, I raise the question about the interrelation between realities. Should we regard ourselves as rooted in a primal natural reality that never basically changes and then see technologies as higher-level realities that remain set into prime reality, just as a Chinese box fits nested inside another Chinese box? Do we enjoy the security of a primary reality that, say, medieval civilization agreed upon? Do we trace our sense of unified reality back to a single God or point of origin? If we do not have a common grounding, then how about finding some existential anchors to ground virtual realities and distinguish them from primary reality?

When the human guinea pig, the boy Jobe, enters with Dr. Angelo the cybernetic chambers of VSI (Virtual Space Industries), he says: "Looks like dungeons in here, Dr. Angelo." And the vaults are indeed dark and ominous, the cybersuits hanging in their gyrospheres like corpses twisted on the rack in a gothic outer space. The difference from the natural healthy body comes out in Dr. Angelo's description of how the cybersuits tap into the endocrine system to stimulate the pituitary, adrenal, and thyroid glands so they work in sync with the audiovisual simulations. VR will ultimately control the electromagnetic currents in the human body and so affect the natural energies and biorhythms of human life. As a teacher and long-time practitioner of the Chinese art of Tai Chi Chuan, I look with anxiety on that kind of control, because I know how delicate human energy systems are and how much internal discipline it takes to integrate and harmonize the system. When VR technology affects our internal energy system—even through a merely audiovisual link—how much inner harmony and mind–body unity can we hope to maintain?

The boy Jobe in *Lawnmower Man* uses computerized "cyberlearning" to accelerate his educational progress. Among other things, he learns classical Latin in two hours. Typical of cyberlearning (and of hypertext thinking) is Jobe's comment as he whips out one compact disc after another, replacing each of them in seconds. "I listen to music by sampling segments," he explains cheerfully. This high-speed learning points out a danger in computerized information access. The world of artificial reality may gradually erode the sense we have of the felt resistance of the real world. Real-world resistance has made us develop a mind-set that contemplates, reflects, and mentally digests matters by chewing them over slowly and thoughtfully. *Festina lente* goes the ancient Roman proverb: "Make haste, but make it slowly" lest it enervate. When he sees the direction in which his VR research is going, Dr. Angelo worries: "This technology can propel the evolution of the human mind, but the technology must be tempered with wisdom." High-speed flipping through great artworks does not bring about wisdom, only more information. Infomania retards rather than accelerates wisdom.

In the final scenes of the movie, as Jobe grows more powerful, he decides to project himself permanently into the computer's animations; he wants to merge with the neural network of computers as a fully electronic entity (like Max Headroom in the earlier film of the same name). By now the boy has taken on messianic airs "to cleanse this diseased planet," and Larry Angelo says of the boy whom he blessed with accelerated learning: "Jobe's insights seem twisted—I fear for Jobe's sanity." Once projected fully into the computer system, Jobe gains control of the virtual world without the limitations of his former physical body. He becomes a cyber entity that rules the virtual world, enjoying the vision of a world of information that has no opaque solids to hide things from his all-probing eyes. Because everything is essentially information, nothing remains hidden from the one who operates the system. As I pointed out in Chapter 7, the model of divine, omniscient Central System Operator directed the seventeenth-century computer research of the philosopher Leibniz. Jobe then draws Larry into cyberspace, suspends him on a crucifix, and pierces Larry's inmost thoughts so they lie open to be read. The boy then exclaims: "I am God here!"

What prevents the virtual-entity Jobe from being completely divine—what preserves his humanity—is the memory of a person he loved as a child when in his former human body. Little Peter, Jobe's young friend, remains a remembered and valued human being in the primary world. With a bomb threatening the body of little Peter, Jobe suspends his omniscient tyranny and commands, "Go save Peter!" And so the bridge between the primary and the virtual world establishes once again the importance of existential care, of personal pain and loss, of limited lifetimes.

The closing scenes spell out a bit more just what values should underpin virtual-worlds research, what wisdom might mean for the computerized world. The members of a potential nuclear family—a female, a child, and the male researcher—overcome together the disaster of the first phase of VR experiment. They stand together, these representatives of primary sensuous life, at the beginning of a second try. The researcher knows he has a second chance, but he must now work underground, outside the establishment. Sheer excitement, thrills, and ecstasy no longer govern his research. "This technology must free the mind of man," he says, "not enslave it." His freedom now includes human companionship. As he stands there, trying to understand the challenge, the phones ring loudly.

All the phones are ringing. They announce the advent of a unifying network of human presence. The electronic world celebrates the moment when circuits coalesced with human beings. The world has unfolded a new dimension, a virtual reality. The phones are indeed ringing. And they ring with metaphysical questions. Who can you call? Looks like it's time again for philosophy!

USEFUL VOCABULARY
FOR THE METAPHYSICS
OF VIRTUAL REALITY

The following definitions offer vocabulary tips for readers who want to understand the metaphysics of virtual reality. The glossary covers the basic language of VR hardware and software, as well as certain terms that illuminate the philosophical issues in VR. With this glossary I am not attempting to exhaust the subject or to preempt the discussion of the issues. Rather, the glossary calls for further debate and future refinements. I offer it as a working guide for the reader who is looking into the field. Any additions or corrections should be sent by e-mail to the author at the Internet address mheim-@beach.csulb.edu or at the WELL at mheim@well.sf.ca.us.

aesthetics   The branch of philosophy investigating questions such as What makes something a work of art? Are there absolute values in art, or are aesthetic values relative? Are aesthetic arguments based on only personal preference? VR realism relies today on traditional aesthetics while promising, because of its power over the user, to reshape the field of aesthetics.

agents   Software objects that perform actions in virtual worlds. They can change, evolve, and learn. The creation of software agents draws on techniques from computer animation, artificial intelligence, neural networks, genetic algorithms, artificial life, and chaos theory.

altered states   An emphasis in psychology focusing on paranormal states of consciousness, such as drug-induced hallucinations, isolation tanks, and religious rapture. Altered-states research interprets VR as a pattern of private perception rather than a real-world connection. As a result, the ontological questions raised by VR go beyond the purview of altered-states research.

analytic/synthetic Whereas analytic thinking takes things apart; synthetic thinking brings things together. Much of early-twentieth-century philosophy emphasized the analytic process and the search for telling details. With the advent of computer support, philosophy has become free for holism, for connections and associations, for ways to assemble illuminating combination. Multimedia, hypertext, and virtual reality express and reinforce the search for synthetic wholes.

artificial life Computerized agents that simulate biological life-forms have artificial life, or a-life. Such agents reproduce, evolve, and carry out the dynamic processes of organic life. Not to be confused with Myron Krueger's artificial reality. See Pattie Maes, ed., *Designing Autonomous Agents: Theory and Practice from Biology to Engineering and Back* (Cambridge, Mass.: MIT Press, 1990).

artificial reality (AR) AR has a precise meaning in the work of Myron Krueger, a pioneer of VR. Krueger's "computer-controlled responsive environment" (p. 10) means an unencumbered involvement in a computerized environment. Computerized sensors "perceive human actions in terms of the body's relationship to a simulated world. The comuter then generates sights, sounds, and other sensations that make the illusion of participating in that world convincing" (p. xii). Artificial reality belongs to VR in the sense that participants aesthetically enter a computer-enhanced environment, but AR systems do not require goggles or datagloves, and such systems involve full-body motion without wiring humans to an interface. The Mandala system and the Cyber Cave from the University of Illinois exemplify recent offshoots of AR. For details, see Myron Krueger, *Artificial Reality*, 2nd ed. (Reading, Mass.: Addison-Wesley, 1991).

augmented reality The superimposition of computer-generated data over the primary visual field. An operating surgeon, for instance, may wear dataglasses to augment the perception of the patient's body, thereby gaining continually updated information on vital signs. The jet fighter's heads-up display is an early form of augmented reality.

AWS (Alternate World Syndrome), AWD (Alternate World Disorder) Flight simulators can cause nausea and disorientation because of the discrepancies between the pilot's actual physical movement (or lack of movement) and the perceived motion in the simulator. The delay between the user's head motion and the computer-simulated motion also adds to the "barfogenic zone." Similarly in VR, a conflict of attention can arise between the cyberbody and the biobody. In this case, an ontological rift appears as the felt world swings out of kilter, not unlike jet lag. In Alternate World Syndrome, images and expectations from an alternate world upset the current world, increasing the likelihood of human errors. If the Alternate World Syndrome (AWS) bevomes chronic, the user suffers Alternate World Disorder (AWD), a more serious rupture of the kinesthetic from the visual senses of self-identity. Treatments for AWS or AWD range from de-linking exercises in cyberspace to more demanding disciplines, such as tai chi and yoga, that restore the integrity of somatic experience.

bandwidth   The amount of data that can be transmitted per second through the lines of an electronic network. VR requires a bandwidth equal to the amount of realism required for the particular application. Realism ranges from photorealism to a softer evocative realism to a sharp-focus–soft-edges realism. Whether VR should strive for realism or for a more imaginative world remains a disputed question in the VR industry. Experiments with different applications will offer answers over time, but a full grasp of the question awaits an in-depth understanding of realism and presence.

BOOM (binocular omni-orientational monitors)   A stereoscopic box mounted on a desktop that floats like a periscope and provides a wraparound three-dimensional imagery (developed by Fake Space Labs).

CAD/CAM (computer-aided design or computer-aided manufacture)   Software for computerizing the industrial-design process so that the initial planning stages of a product have greater precision and flexibility than when drawn by hand. The two-dimensional screen of CAD/CAM software limits the user to perspectival drawing that only suggests three dimensions. The CAD software company Autodesk, the creator of AutoCAD(tm) software, leads in developing virtual reality and cyberspace software with "6 degrees of freedom" (x, y; and z axes, plus the three orientations of roll, pitch, and yaw).

CompuServe   A national computer network that operates commercially to link hundreds of thousands of personal computers. CompuServe has a virtual-reality forum in the COMART (Computer Art) section, Room 13, moderated by John Eagan, who can be contacted for details at CompuServe ID no. 76130,2225.

cyber   A prefix found throughout the literature of VR. The root reference is to *cybernetics*, the science of self-regulating systems, but the reference has expanded to become a name for mainframe computers (the Cyber 960) and now connotes the human involvement with computers (the cyborg, or cybernetic organism). For instance, the primary human body becomes a cyberbody when appearing in the cyberspace of a virtual environment.

CyberEdge Journal   A source magazine for VR. covering virtual reality and related subjects. For information, contact the editor, Ben Delaney, at bdel@well.sf.ca.us.

cyberglove   Another form of the dataglove, or device for monitoring hand movements so that the user's position and gestures can be calculated and the computer can adjust the graphic virtual environment accordingly.

CyberJobe   The science fiction character Jobe Smith in the movie *Lawnmower Man*. CyberJobe is a twenty-first-century Frankenstein. In this cautionary tale, a young half-wit grows intelligent as well as menacing once he is psychologically altered by visits to an experimental cyberspace and virtual reality. The movie's computerized special effects overwhelm any verbal and humanistic content the story may have had.

cybernation   The application of computers and automatic machinery to carry out complex operations. Managers in government and industy cybernate many complex, repetitive tasks by introducing computers.

cyberpunk   A postmodern literary–cultural style that projects a computerized future. The future is dominated by private corporations that use information technology and drugs to control individuals. Cyberpunk stories are told from the criminal perspective and portray the widespread use of biotechnology, computers, drugs, and a paranoid life-style. Individuals increasingly merge with electronic devices, and hallucinations rule public life. Cyberpunk is based on a dystopian brand of science fiction whose patron saint is Philip K. Dick and whose manifesto is William Gibson's novel *Neuromancer*. The term was coined by the science fiction writer Bruce Bepkie and became a literary critical term with Gardner Dozois, the editor of *Isaac Asimov's Science Fiction Magazine*.

cyberspace   The juncture of digital information and human perception, the "matrix" of civilization where banks exchange money (credit) and information seekers navigate layers of data stored and represented in virtual space. Buildings in cyberspace may have more dimensions than physical buildings do, and cyberspace may reflect different laws of existence. It has been said that cyberspace is where you are when you are having a phone conversation or where your ATM money exists. It is where electronic mail travels, and it resembles the Toontown in the movie *Roger Rabbit*.

dataglove, cyberglove   A sensor-laced nylon glove that provides manual access to objects in virtual encironments, sometimes also enabling a variety of gestures to initiate movements in the virtual world. That is, the glove has fiber-optic sensors to track hand and finger positions, permitting the user to reach out, grab, and change objects in the virtual world. Present-day datagloves register the hand positions and the degree of movement of each finger but do not yet register the somatics, the inward kinesthetics of movement that are more difficult to measure.

datasuit   A sensor-equipped garment like the dataglove but covering the whole body in order to track the user's movements and to provide constant input into the host computer so the computer-generated graphics environment and the cyberbody can be updated according to the user's gestures and orientation.

determinism   The view that every event occurs necessarily, follow-ing inevitably from the events preceding it. Determinism rejects any randomness, considering freedom to be either an illusion (hard deter-minism) or subject in some way to necessity (soft determinism). Soft determinism might be applied to technological change in the follow-ing way: the introduction of a technology inevitably transforms soci-ety; as a working individual in society, I have no real choice but to participate in implementing the new technology; the manner in which I implement the technology, however, can be thoughtful or reflective, critical or conservative, depending on the attitude I consciously adopt. Hence, the how of my individual response is nei-ther reactionary (rejecting technology wholesale) nor utopian (herald-ing technology as a panacea). Such technological determinism is a variant of soft determinism.

dualism   The view, stemming from the seventeenth-century philoso-pher René Descartes, that mind (thinking substance) can work on its own, apart from matter (extended substance), to constitute a full-ness of reality. According to this view, matter exists only in inert substances and cannot be understood outside the sciences. Descartes founded analytic geometry and contributed to geometrical optics in his *Discourse on Method* (1637). Cartesian philosophy continues to have an influence on the debate about VR realism.

e-mail, electronic mail   Messages delivered by networked computers. An e-mail network may be local within a single corporation, or it may extend to hundreds of nodes throughout the world, such as In-ternet. Users receive and answer mail messages at a terminal or per-sonal computer, and the system relays the messages in seconds to the address where they arrive for the receivers to view at any conve-nient time. Such systems are asynchronous rather than real time because the users need not be present at the time of delivery.

enhanced virtual vision   Continuous three-dimensional data super-imposed over video images and pumped across a workstation into eyephones helps locate missing graphic elements or disparities. It is similar to augmented reality but without the direct contact with the primary visual world.

entity   Something that registers as ontologically present or that has an effect on a world. Entities include virtual objects as well as expert systems functioning as agents in the virtual world. Entities need not reflect real-world metaphysics but may draw on imaginative and/or spiritual traditions, such as the Loas of Voodoo (Gibson) or the myth systems of the great religions (Rogers's "Mythseeker").

epistemology The traditional study of human knowledge, its sources, its validation, and its implications for other aspects of life.

existential, existentialism A philosophical emphasis on presence and making present, on action and human choice. Existentialism was a movement in early-twentieth-century philosophy that rejected static, essentialist worldviews in favor of pragmatic, risk-taking encounters with history and change. Its main proponents were Martin Heidegger, Jean-Paul Sartre, Simone de Beauvoir, and Karl Jaspers. Philosophers like Maurice Merleau-Ponty emphasized the role of the human body in achieving presence. The interactivism of the late twentieth century carries on certain themes of existentialism.

eyephones A head-mounted display that links the user's visual field with the computer-generated images of a virtual world. Eyephones shut out the primary visual world and supply the user with a continuous stream of computer-generated three-dimensional images. A variety of devices, from stereoscopic booms to virtual retinal displays, feed input into the user's optic sense.

eye-scanning devices The host computer needs information about the user's eye movements in order to generate constantly and update appropriate visual images. A variety of devices, from head tracking to low-level lasers, are used to supply and compute the user's visual activity. As Aristotle pointed out in *Metaphysics* over two thousand years ago, humans have tended to learn more about the outside world from their sense of vision than from any of their other senses, because eyes deliver the most detailed and differentiated field of information. Existentialist philosophy, however, has meanwhile challenged the primacy of vision by pointing out the importance of presence and the nonvisual cues that alert us to presence.

flight simulator A precursor of VR that emerged before World War II as an aid to pilot training. The earliest simulators used photographs coupled with motion machines to imitate the feel of flying an aircraft.

glass-bead game A fictional game described by Hermann Hesse's novel *Das Glasperlenspiel* (1943), translated in English as *Magister Ludi* (the game master). Discussions of VR often evoke references to the glass-bead game because the game's players combine all the symbols of world cultures so as to devise surprising configurations that convey novel insights. Each player organizes the cultural symbols somewhat like a musician improvising on an organ that can mimic any instrument. The glass-bead game's synthetic, non-linear information play is a forerunner of hypertext and of virtual worlds. Hesse's fiction also touches on some of the human problems underlying the advent of cyberspace and virtual reality, such as the role of the body and of disciplines for deepening the human spirit.

GUI (graphical user interface)   A term used by the computer industry to distinguish one specific approach to interacting with a computer. GUI is distinguished from the Command Line Language (CML). The alphanumerics of a computer's command line (such as the C prompt on MS-DOS personal computers) present more abstract, less body-embedded symbols than GUI does (such as on Apple Macintosh computers). GUI dramatizes the user's bodily action by making the "delete" command, for example, into the motion of carrying files (with a mouse) out to the (icon of a) trash can.

HCI (human–computer interaction)   HCI studies how to access computer power. The computer industry developed "human factors engineering" to explore different kinds of input ranging from binary code to alphanumeric keyboards to touchscreens and the mouse or track ball. Seen from the HCI viewpoint, VR is a latest development in "user friendlimess." The HCI approach, however, misses the broader implications of the human entrance into virtual reality.

HDTV (high-definition television)   HDTV advances the bandwidth of photo representation to achieve a higher degree of visual realism. Television lacks the interactive and immersive features of VR.

head-mounted display (HMD)   Also known as virtual-reality goggles of eyephones or, sardonically, "the Facesucker." The device covers both eyes and renders real-time stereoscopic graphics generated by the host computer. The HMD also provides head-tracking information so that the computer can generate perspectives of the virtual world appropriate to the user's bodily orientation and head/eye movements. Some HMDs also come equipped with headphones for audio tracking of virtual entities. "Visette" is the British term for the HMD manufactured by W Industries of the United Kingdom. More advanced techniques include the use of low-level laser lights to send beams directly onto the retina, creating holographic representations.

heads-up display (HUD)   A visor that, combined with head-positioning sensors, augments the visual field of the user. The visor superimposes a virtual floating window display that acts as an electronic associate, providing image projections of assembly instructions or blueprints to guide workers during their manufacturing tasks.

Holodeck   An idealized computer-to-human interface from the

Vocabulary for the Metaphysics of Virtual Reality

science fiction television series "Star Trek: The Next Generation." The Holodeck is a room where spoken commands call up images in realistic landscapes populated with walking, talking "humans" (artificial personalities) and detailed artifacts that appear so lifelike that they are indistinguishable from reality. Used by the crew of the starship *Enterprise* to re-create and to visit such times and places as medieval England and 1920s America.

hyper   The prefix *hyper* means "extended." *Hyperspace* is space extending beyond three dimensions. *Hypersystems* are nonlinear linked systems in which one link may route directly to a link on an entirely different plane or dimension. *Hypermedia* cross link information in text, graphics, audio, or video.

hypertext   An approach to navigating information. From the computer science point of view, hypertext is a database with nodes (screens) connected with links (mechanical connections) and link icons (to designate where the links exist in the text). The semantics of hypertext allows the user to link text freely with audio and video, which leads to hypermedia, a multimedia approach to information. Many prototype hypertext systems, such as KNS, IBIS, INTERMEDIA, and NOTECARDS, used the model of computer index cards (which later became Hypertext "Stacks") containing data such as text, graphics, and/or images. In 1982, the ZOG hypertext was installed as a computer-assisted information mamagement system on the USS *Carl Vinson*, a nuclear-powered aircraft carrier. ZOG is the most fully tested hypertext system to date. Its hypertext features include filtering through points of view or subjects, aggregative structures rather than shared logical structure, and versioning that keeps the history of the modifications to the database. The term *hypertext* was coined by Ted Nelson in 1964. The disadvantages of hypertext include disorientation and cognitive overload.

IA (as opposed to **AI**) (**intelligence amplification**)   The sum of the interface of human and computer. The human being intuits patterns, relations, and values, and the computer processes and generates data that include the sensory input for human senses. IA replaces the AI obsession with human–computer rivalry by seeing the codependence of human and computer.

idealism   The ontological view that in the final analysis, every existing thing can be shown to be mental or spiritual. In the West, this view is usually associated with the views of George Berkeley and Georg Hegel.

immersion   An important feature of VR systems. The virtual environment submerges the user in the sights and sounds and tactility specific to that environment. Immersion creates the sense of being present in a virtual world, a sense that goes beyond physical input and output. How presence and immersion coalesce remains an open question in VR research.

infonaut, cybernaut Terms used to describe those who "swim" among virtual polymer molecules, who enter the virtual interior of galactic black holes, and who explore scientific data by moving through their representations in a virtual world.

input Information supplied to a computer through a variety of different devices like keyboards, a mouse, a joystick, voice recognition, and position trackers such as helmets, BOOMs, and datagloves.

interface The locus of communication between two systems, applied to either hardware or software or a combination of both. A graphical interface, for example, may use metaphors such as a desktop or house with garbage pail, paintbrush, or yardstick. An alphanumeric interface, such as that of an IBM-style personal computer, consists of a monitor and a keyboard and the appropriate software for input and output. Interface is a key term in the philosophy of technology because it designates the connecting point between human and digital machine. See Chapter 6.

irrealism *World* is a plural concept, according to Nelson Goodman in *Ways of Worldmaking*. Each world is a variant of related worlds, and each world makes its own context and rules of intelligibility. There is a world of sports and a world of art and a world of religion and a world of science. There are also times and places when we have a need for making ourselves present in a single world. This is irreal because it undermines the uncritical affirmation of a single world. Irrealism parallels Heidegger's existentialist notion of world in *Being and Time* (1927).

materialism The ontological view that in the final analysis, everything can be shown to be material and that mental and spiritual phenomena either are nonexistent or have no existence independent of matter. In the West, this view is usually associated with Democritus, Thomas Hobbes, and Karl Marx.

metaphysics, metaphysical The study of the first principles of reality, including speculation on epistemology (knowledge), ontology (being), ethics (goodness), and aesthetics (beauty). The metaphysics of VR treats issues such as presence, degrees of reality, objectification (first person, third person). simulation versus reality, the ratio of mental to sensory material, the ethics of simulation, the evaluation of virtual environments, and the central coordination of virtual realities. Traditional metaphysics also treats topics such as possible worlds, intrinsic goals (teleology), and umbrella concepts like meaning and final purpose.

mirror worlds A software concept developed by David Gelernter at Yale in which the computer creates real-time minature maps that mirror the larger world in which the user is present. Mirror worlds appear in some measure on the bird's-eye view of the jet pilot's radar screen, and they were also foreshadowed earlier by prototypes in ancient philosophy and art.

mobility devices   Stationary bikes, track balls, flying mice, treadmills, and other hardware connecting human motions with computer-generated environments.

network, the net, the matrix   A network connecting computers through cables, telephone lines, or satellite transmission. The global Internet network connects institutions of all kinds: military and government, commercial and educational. Networks also exist in local areas such as a business and on commercial mainframe computers, such as those used by CompuServe and Prodigy. Most often, gateways exist through which one network opens onto another.

neutral interface   Projected by science fiction, the notion of connecting human–computer input and output by tapping immediately into the nervous system of the human user. Discouraged by neurologists because of its obvious dangers, the notion still persists among scientists looking for an "ultimate interface."

nihilism, nihilistic   The view that either nothing truly exists or nothing deserves to exist.

object-oriented programming (OOP)   Programming languages such as C and C+ + offer solutions to the complex problem of programming virtual worlds. Each situation in a virtual world requires a complex logical calculus. For example, if X number of agents exist in a world and all can have one interaction with one another, then the interactions will multiply to $X ' 2 - X$. If a variety of interactions are possible, say 9, then 10 agents will require 90 interactions; 100 agents, 9.900; and so on. OOP reduces the exponential growth of software complexity. The qualities of OOP languages—encapsulation, association, and polymorphism—reduce the complexity of programming virtual worlds, since multiple interactions do not have to be programmed individually.

ontology, ontological   The study of the relative reality of things. An ontology ranks some things as "more real" or "actually existing," as opposed what is unreal, phony, fadish, illusory, ephemeral, or purely perceptual. Ontology locates the difference between real and unreal and then develops the implications of that way of differentiating the real from the unreal. Traditional ontology studies entities or beings by observing the conditions under which we ascribe reality to beings. Ontology in the existential sense goes beyond traditional ontology by noticing the holistic background against which entities appear. The existential "world" in which entities appear also changes over time. The ontological shift constitutes a change of context according to which the realness of entities must be recalibrated. Epistemology, the study of knowledge, takes its bearings from clearly known entities and so does not dig down to the field of ontology. Because knowledge is based on assertions and propositions, the results of epistemology can stand up to greater logical scrutiny. Ontology, on the contrary, relies more on intuitive awareness and peripheral noticing than on argument and logic—which is not to say that ontology rejects argument and logic.

pragmatic   A classical position in philosophy emphasizing the point of view of the user. The important questions are not separable from questions of human action. Pragmatism is the metaphysics of human factors.

presence   A notion crucial to early-twentieth-century philosophy. In Heidegger's *Being and Time* (1927), presence is synonymous with being and is a function of temporality. The entire history of reality, according to Heidegger, must be reconsidered from the standpoint of presence. *Presence* is also a key term in VR, with researchers seeking to define and quantify the presence that a given system will deliver.

primary world   The world (context of human involvements) outside the computer-generated world. The primary world has distinguishing properties such as natality or mortality, fragility or vulnerability to pain and injury, and personal care. The ontology of the primary world limits the realism of the virtual world.

psychological atomism   The view that all knowledge is built from simple, discrete psychological data, such as primitive sensorial experiences of colors, sounds, and tastes. In the West, this view is usually associated with John Locke, George Berkeley, and David Hume.

realism   In a technical sense, realism refers to metaphysical theories that attribute priority to abstract entites. Platonism, for instance, is a kind of realism in maintaining that mathematical patterns are more real than are their instances in the physical world. Platonism finds the reality of things in their stability, intelligibility, and reliability for the knower. In a related sense, realism is the approach that treats cyberspace as an actual (phenomenological) world with its own particular kind of entities. Nonrealists approach cyberspace and VR as hardware and/or software configurations separate from the user's experience. Realists speak of the net and the matrix as actual places.

real time   Simultaneity in the occurrence and registering of an event, sometimes called *synchronous processing*, as opposed to *asynchronous processing*, in which the event remains at a distance from its registration as data.

robotic telepresence   The science of driving robots remotely and attaching video cameras to them to perform building inspection and other tasks with a human user's presence. Robotization has begun in underwater and space construction as well as in the handling of nuclear waste.

SIMNET A three-dimensional visual simulator developed for virtual-world exploration in the military. SIMNET is distributed interactive simulation (DIS) linking numerous simulation sites at one time so that the users can interact with users at other sites. SIMNET originally helped trained tank combat crews practicing communication techniques. One team drives its tank over a simulated terrain and encounters the tank driven by another crew on a remote or near site.

Developed at the Naval Postgraduate School, NPSNET is an offshoot of SIMNET. It allows the participant to select a vehicle by means of a button box and to drive the vehicle over the ground or in the air in real time with a space ball that allows control with 6 degrees of freedom. The displays show on-ground cultural features such as roads, buildings, soil types, and elevations. It supports a full complement of vehicles, houses, trees, signs, water towers, and cows; and it can represent environmental effects such as San Francisco–like fog and Los Angeles–like smog.

solipsism The view that the only true knowledge one can possess is the knowledge of one's own consciousness. Solipsism maintains that there is no good reason to believe that anything exists other than oneself.

somatics, somatic An inside view of your own body taken from the first-person point of view, as opposed to a third-person point of view that looks at the body from the outside. The term *somatics* comes from the Greek *soma*, which means "body," and derives from Thomas Hanna's studies of integral body experience. Non-Western approaches to health and medicine, including Yoga and Tai Chi, begin with somatic assumptions. Terms such as *energy* (*chi*, *ki*, and *prana*) gain their phenomenological meaning from the application of attention to the sensation of first-person physical processes. Western medicine insists on a split between the observer and the observed, the mind and the body, whereas the Eastern approach seeks a fuller presence that harmonizes the mind with the body.

spacemaker A designer of cyberspace constructs, like a filmmaker. The term was first introduced at Autodesk to indicate the unity of design and construction, because to plan something in cyberspace is tantamount to building it.

substance A traditional term referring to what is considered to be the most basic, independent reality. For Aristotle, the color of a horse is not a substance because the color cannot exist independently, but the horse is a substance because it can exist independently (of its color or size). For Spinoza, only God can be said to truly exist because only God is a completely independent being. George Berkeley believed that material things cannot exist independently of perception, and Immanuel Kant relegated substance to a category of human thinking.

Taoism, Taoist   A stream of culture emerging in ancient China. Taoism conceives Nature as a continual balancing of yin/yang energies. More important than the theory was the practice of Taoism. Taoists invented hundreds of practices, from acupuncture to Tai Chi movement, from meditation to healing and artistic skills. These ancient practices still carry Taoism throughout the world, always on the periphery of the Western scientific conception of the human being.

teleology, teleological   An explanation in terms of goals or purposes. The intention may be distant (Greek, *telos*) from the action, but the intention informs the action, giving it teleological reality. Teleological thinkers tend to equate reality and meaning.

telepresence   Operations carried out remotely while the user remains immersed in a simulation of the remote location. (The Greek word *têle* means "at a distance," and so *telepresence* means "presence at a distance.") Robotic mechanisms make telepresence effective at a remote site. For example, telepresence surgery allows surgeons to combine robotic instruments with endoscopy (cameras inserted in the patient's body). NASA uses slightly asynchronous transmissions to achieve telepresence outside terrestrial space.

telerobotics   The technique of using mechanical devices to do work (Russian, *robotayu*) at a distance. Add the cybernetics of information systems, and the telerobot becomes a component of telepresence.

three-dimensional sound   Sound reproduction on the space of a virtual world occurs at every digital point, creating a sense of precise source location. Sounds seem to occur above, below, in front of, and to either side of the listener. Such omnidirectional sound produces an "acoustic photograph" of a sound environment. VR three-dimensional sound has emerged with advances in psychoacoustic research combined with the development of digital-signal processing. Sound augmentation, similar to the augmentation of visual reality, uses computers to overlay and enhance the sound space of entire areas, like the subway in Berlin, Germany.

trackers   Position-tracking devices that constantly monitor the user's physical body motions—hand, head, or eye movements—so as to feed the user's actions into the host computer, in which motions are interpreted as changes in the computer-generated environment. Some of the earliest devices for tracking head position, and hence visual perspective, are the 3Space Isotrak by Polhemus, Inc., and the Bird by Ascension Technology. Position-tracking devices by themselves do not register the user's somatic states.

virtual   A philosophical term meaning "not actually, but just as if." It came into recent vogue with the use of computer techniques to enhance a computer memory. Virtual-memory techniques extend the data storage of a computer without adding hardware. On a personal computer, for example, virtual memory can be a part of RAM used as though it were a hard disk storage space. Such a virtual disk can be used like a hard disk, but does not have the physical limitations of an actual mechanical disk. Similarly, something can be present in virtual reality without its usual physical limitations. The debate about the value of virtual existence has appeared throughout the history of philosophy, with an especially vigorous debate in the era of Duns Scotus at the end of the medieval period and the beginning of the age of nominalism.

virtual reality (VR)   Virtual reality pertains to convincing the participant that he or she is actually in another place, by substituting the normal sensory input received by the participant with information produced by a computer. This is usually done through three-dimensional graphics and input–output devices that closely resemble the participant's normal interface with the physical world. The most common I–o devices are gloves, which transmit information about the participant's hand (position, orientation, and finger bend angles), and head-mounted displays, which give the user a stereoscopic view of the virtual world via two computer-controlled display screens, as well as providing something on which to mount a position/orientation tracker.

The definition of VR includes several factors and emphases: artificial reality, as when the user's full-body actions combine with computer-generated images to forge a single presence; interactivity, as when the user enters a building by means of a mouse traveling on a screen; immersion, as when the user dons a head-mounted display enabling a view of a three-dimensional animated world; networked environments, in which several people can enter a virtual world at the same time; telepresence, in which the user feels present in a virtual world while robotic machines effect the user's agency at a remote location in the actual primary world. See Chapter 8.

virtual world, virtual environment   A scene or an experience with which a participant can interact by using computer-controlled input–output devices. Most virtual worlds attempt to resemble physical reality, but controversy continues about the value of various levels of resemblance. Virtual worlds are not tied to physical reality, since any information that can be visualized can also be made into a virtual world that a participant can experience. Cyberspace, in other words, contains many kinds of virtual worlds. Even if a virtual world imitates a physical world, a decision has to be made whether the VR should imitate the perceived world of human phenomena or the world known to physical sciences (which often defies the assumptions of human phenomenology).

world A world, either virtual or real, is a total environment for human involvement, such as the "world of sports" or the "world of the Otavalo Indians," or the "world of nuclear physics." The world in a singular sense refers to the horizon or totality of all involvements. A virtual world represents things so that they have an artificial presence. To prune exaggerated expectations, research laboratories often prefer to speak of virtual worlds rather than virtual reality (Autodesk) or virtual environments rather than virtual worlds (MIT).

SELECTED READINGS

Abramson, Jeffrey B., F. Christopher Arterton, and Gary R. Orren. *The Electronic Commonwealth: The Impact of New Media Technologies on Democratic Politics.* New York: Basic Books, 1988.

Agosti, Maristella. "Is Hypertext a New Model of Information Retrieval?" In *Proceedings of the 12th International Online Information Meeting: December 1988, London, England.* Medford, N.J.: Learned Information, 1988.

Alexander, Frederick Matthias. *The Resurrection of the Body: The Writings of F. M. Alexander.* Ed. Edward Maisel, with a preface by John Dewey. New York: University Books, 1969.

Austakalnis, Steve, and David Blatner. *Silicon Mirage: The Art and Science of Virtual Reality.* Berkeley, Calif.: Peachpit, 1992.

Barrett, Edward, ed. *The Society of Text: Hypertext, Hypermedia and the Social Construction of Information.* Cambridge, Mass.: MIT Press, 1989.

———. *Text, Context and Hypertext: Writing with and for the Computer.* Cambridge, Mass.: MIT Press, 1988.

Basalla, George. *The Evolution of Technology.* Cambridge: Cambridge University Press, 1988.

Bateson, Gregory. *Steps to an Ecology of Mind.* San Francisco: Chandler, 1972.

Baudrillard, Jean. *Simulations.* Trans. Paul Foss, Paul Patton, and Philip Beitchman. New York: Semiotext(e), 1983.

Benedikt, Michael, ed. *Cyberspace: First Steps.* Cambridge, Mass.: MIT Press, 1991.

Benjamin, Walter. "The Work of Art in the Age of Mechanical Reproduction." In *Illuminations,* ed. Hannah Arendt. New York: Schocken Books, 1969.

Berk, Emily, and Joseph Devlin, eds. *The Hypertext/Hypermedia Handbook.* New York: McGraw-Hill, 1991.

Berkeley, George. *An Essay Towards a New Theory of Vision*. London: Dent, 1709.

Berman, Morris. *Coming to Our Senses: Body and Spirit in the Hidden History of the West*. New York: Bantam Books, 1990.

Bevilacqua, Ann F. "Hypertext: Behind the Hype." *American Libraries*, February 1989, pp. 158–62.

Bijker, Wiebe, Thomas Hughes, and Trevor Pinch. *The Social Construction of Technological Systems*. Cambridge, Mass.: MIT Press, 1987.

Bolter, Jay David. "The Idea of Literature in the Electronic Age." *Topic: A Journal of the Liberal Arts* 39 (1985): 23–34.

———. "Topographic Writing: Hypertext and the Electronic Writing Space." In *Hypermedia and Literary Studies*, ed. Paul Delany and George Landow. Cambridge, Mass.: MIT Press, 1991.

———. *Writing Space: The Computer, Hypertext, and the History of Writing*. Fairlawn, N.J.: Erlbaum, 1990.

Borchmeyer, Dieter. *Richard Wagner: Theory and Theater*. Trans. Stewart Spencer. Oxford: Clarendon Press, 1991.

Borges, Jorge Luis. *Labyrinths*. New York: New Directions, 1964.

Borgmann, Albert. *Crossing the Postmodern Divide*. Chicago: University of Chicago Press, 1992.

———. *Technology and the Character of Contemporary Life: A Philosophical Inquiry*. Chicago: University of Chicago Press, 1984.

Brand, Stewart. *The Media Lab: Inventing the Future at MIT*. New York: Viking Penguin, 1987.

Brod, Craig. *Technostress: The Human Cost of the Computer Revolution*. Reading, Mass.: Addison-Wesley, 1984.

Brooks, Frederick Phillips, Ming Ouh-Young, James J. Batter, and P. Jerome Kilpatrick. "Grasping Reality Through Illusion: Interactive Graphics Serving Science." In *Proceedings of the Computer Human Interaction Special Interest Group of ACM, 1988*. Washington, D.C., May 15–19, 1988.

Bush, Vannevar. "As We May Think." *Atlantic Monthly*, July 1945, pp. 106–7.

———. *Science Is Not Enough*. New York: Morrow, 1967.

Campbell, Jeremy. *Grammatical Man: Information, Entropy, Language, and Life*. New York: Simon and Schuster, 1982.

Carrington, Patricia. *Freedom in Meditation*. New York: Doubleday, 1977. Reprint. Kendall Park, N.J.: Pace Educational Systems, 1984.

Daiute, Colette. *Writing and Computers*. Reading, Mass.: Addison-Wesley, 1985.

Delany, Paul, and George Landow, eds. *Hypermedia and Literary Studies*. Cambridge, Mass.: MIT Press, 1991.

DeLillo, Don. *White Noise*. New York: Penguin, 1985.

Douglas, Jane Yellowlees. "Wandering Through the Labyrinth: Encountering Interactive Fiction." *Computers and Composition* 6 (1989): 93–103.

Drexler, E. *Engines of Creation: Challenges and Choices of the Last Technological Revolution*. Garden City, N.Y.: Doubleday, 1986.

Dreyfus, Hubert. *Mind over Machine: The Power of Human Intuition and Expertise in the Era of the Computer*. New York: Free Press, 1985.

Dunlop, Charles, and Rob Kling, eds. *Computerization & Controversy*. San Diego: Academic Press, 1991.

Edwards, Deborah M., and Lynda Hardman. "Lost in Hyperspace: Cognitive Mapping and Navigation in a Hypertext Environment." In *Hypertext: Theory into Practice*, ed. Ray McAleese. Norwood, N.J.: Ablex, 1989.

Eisenstein, Elizabeth L. *The Printing Press as an Agent of Change: Communications and Cultural Transformations in Early-Modern Europe*. 2 vols. Cambridge: Cambridge University Press, 1979.

Ellis, Stephen R., Mary K. Kaiser, and Arthur C. Grunwald, eds. *Pictorial Communication in Virtual and Real Environments*. New York: Taylor & Francis, 1991.

Engelbart, Douglas C. "A Conceptual Framework for the Augmentation of Man's Intellect." In *Vistas in Information Handling, Volume I*, ed. Paul W. Howerton and David C. Weeks. Washington, D.C.: Spartan Books, 1963.

Fedorov, Nikolai Fedorovich. *What Was Man Created For? The Philosophy of the Common Task*. Trans. Elisabeth Koutaissoff. London: Honeyglen, 1990.

Fisher, Scott, and Jane Tazelaar. "Living in a Virtual World." *Byte*, July 1990, pp. 215–20.

———. "Virtual Environments: Personal Simulations & Telepresence." In *Virtual Reality: Theory, Practice, and Promise*, ed. Sandra K. Helsel and Judith Paris Roth. Westport, Conn.: Meckler, 1991.

Flim, Leona. "Bookish Versus Electronic Text: Ivan Illich and Michael Heim." M.A. thesis, University of Calgary, 1991.

Fluegelman, Andrew, and Jeremy J. Hewes. *Writing in the Computer Age: Word Processing Skills and Style for Every Writer*. Garden City, N.Y.: Doubleday, 1983.

Friedhoff, Richard, and William Benson. *Visualization: The Second Computer Revolution*. New York: Abrams, 1990.

Furness, Thomas A. "Exploring Virtual Worlds." Interview. *Communications of the ACM*, July 1991, pp. 1–2.

―――. "Virtual Interface Technology (Virtual Reality)." Paper for University of California at Los Angeles, University Extension Program Short Course, 1992.

Garson, Barbara. *The Electronic Sweatshop*. New York: Simon and Schuster, 1988.

Gelernter, David. *Mirror Worlds*. New York: Oxford University Press, 1991.

Gibson, James Jerome. *The Ecological Approach to Visual Perception*. Boston: Houghton Mifflin, 1979.

―――. *Reasons for Realism*. Hillsdale, N.J.: Erlbaum, 1982.

Gibson, William. *Count Zero*. New York: Ace Books, 1986.

―――. *Mona Lisa Overdrive*. New York: Bantam Books, 1988.

―――. *Neuromancer*. New York: Ace Books, 1984.

Goodman, Nelson. *Ways of Worldmaking*. Indianapolis: Hackett, 1978.

Greenberg, Donald P. "Computers and Architecture." *Scientific American*, February 1991, pp. 104–9.

Greenfield, Patricia Marks. *Mind and Media: The Effects of Television, Video Games, and Computers*. Cambridge, Mass.: Harvard University Press, 1984.

Gumpert, Gary, and Robert Cathcard, eds. *Intermedia: Interpersonal Communication in a Media World*. 3rd ed. New York: Oxford University Press, 1986.

Habermas, Jurgen. *The Theory of Communicative Action*. Trans. Thomas McCarthy. Boston: Beacon Press, 1984.

Hanna, Thomas. *Bodies in Revolt: A Primer in Somatic Thinking*. New York: Holt, Rinehart and Winston, 1970.

―――. *The Body of Life*. New York: Knopf, 1980.

―――. *Somatics: Reawakening the Mind's Control of Movement, Flexibility, and Health*. Reading, Mass.: Addison-Wesley, 1988.

Hardison, O. B. *Disappearing Through the Skylight: Culture and Technology in the Twentieth Century*. New York: Viking, 1989.

Havelock, Eric. *Preface to Plato*. Cambridge, Mass.: Harvard University Press, 1963.

Heidegger, Martin. *Basic Writings*. Ed. David Krell. New York: Harper & Row, 1977.

―――. "Hebel—Friend of the House." Trans. Bruce Foltz and Michael Heim. In *Contemporary German Philosophy*, vol. 3, ed. Darrel E. Christensen. University Park: Pennsylvania State University Press, 1983.

―――. *The Metaphysical Foundations of Logic*. Trans. Michael Heim. Bloomington: Indiana University Press, 1984.

―――. *The Question Concerning Technology and Other Essays*. Trans. William Lovitt. New York: Harper & Row, 1977.

―――. *Wegmarken*. Frankfurt: Klostermann, 1967.

Heim, Michael. "The Computer as Component: Heidegger and McLuhan." *Philosophy and Literature*, October 1992, pp. 33–44.

———. "Cybersage Does Tai Chi." In *Falling in Love with Wisdom*, ed. David Karnos and Robert Shoemaker. New York: Oxford University Press, 1993.

———. *Electric Language: A Philosophical Study of Word Processing*. New Haven, Conn.: Yale University Press, 1989.

———. "Grassi's Experiment: The Renaissance Through Phenomenology." *Research in Phenomenology* 18 (1988): 233–63.

———. "Remembering the Body Temple." *Healing Tao Journal* 1, no. 4 (1991): 10–12.

———. "Searching for the Essence of Tai Chi." *Healing Tao Journal* 1, no. 2 (1989): 21–24.

———. "The Sound of Being's Body." *Man and World: An International Philosophical Review* 21 (1988): 469–82.

———. "The Technological Crisis of Rhetoric." *Philosophy and Rhetoric* 21, no. 1 (1988): 48–59.

Helsel, Sandra K., and Judith Paris Roth, eds. *Virtual Reality: Theory, Practice, and Promise*. Westport, Conn.: Meckler, 1991.

Henderson, Joseph. "Designing Realities: Interactive Media, Virtual Realities, and Cyberspace." In *Virtual Reality: Theory, Practice, and Promise*, ed. Sandra K. Helsel and Judith Paris Roth. Westport, Conn.: Meckler, 1991.

Hofstadter, Douglas. *Gödel, Escher, Bach: An Eternal Golden Braid*. New York: Basic Books, 1979.

Horn, Robert E. *Mapping Hypertext: The Analysis, Organization and Display of Knowledge for the Next Generation of On-Line Text and Graphics*. Waltham, Mass.: Lexington Institute, 1990.

Ihde, Don. *Technology and the Lifeworld: From Garden to Earth*. Bloomington: Indiana University Press, 1990.

Illich, Ivan, and Barry Sanders. *ABC: The Alphabetization of the Popular Mind*. San Francisco: North Point Press, 1988.

James, William. *Psychology, Briefer Course*. New York: Collier-Macmillan, 1972.

Johnson, Mark. *The Body in the Mind: The Bodily Basis of Meaning, Imagination, and Reason*. Chicago: University of Chicago Press, 1974.

Joyce, Michael. "Selfish Interaction: Subversive Texts and the Multiple Novel." In *The Hypertext/Hypermedia Handbook*, ed. Emily Berk and Joseph Devlin. New York: McGraw-Hill, 1991.

———. "Siren Shapes: Exploratory and Constructive Hypertexts." *Academic Computing*, November 1988, pp. 10–40.

Kant, Immanuel. *Observations on the Feeling of the Beautiful and Sub-*

lime. Trans. J. T. Goldthwait. Berkeley: University of California Press, 1960.

Kay, Alan. *Personal Dynamic Media*. Xerox PARC Technical Report SSL-76-1. Palo Alto, Calif.: Xerox Palo Alto Research Center, 1976.

Kernan, Alvin. *The Death of Literature*. New Haven, Conn.: Yale University Press, 1990.

Koehler, Ludmila. *Fedorov: The Philosophy of Action*. Pittsburgh: Institute for the Human Sciences, 1979.

Krueger, Myron W. *Artificial Reality*. Reading, Mass.: Addison-Wesley, 1983.

————. *Artificial Reality II*. Reading, Mass.: Addison-Wesley, 1991.

Lambert, Steve, and Suzanne Ropiequet, eds. *CD ROM: The New Papyrus*. Redmond, Wash.: Microsoft Press, 1986.

Lanier, Jaron. "An Insider's View of the Future of Virtual Reality." Interview with Frank Biocca. *Journal of Communication*, Autumn 1992, pp. 150–72.

————. Interview with Stephen Porter. *Computer Graphics World*, April 1992, pp. 61–70.

Laurel, Brenda. *Art of Human–Computer Interface*. Reading, Mass.: Addison-Wesley, 1990.

————. *Computers as Theatre*. Reading, Mass.: Addison-Wesley, 1991.

Levin, David Michael. *The Body's Recollection of Being: Phenomenological Psychology and the Deconstruction of Nihilism*. London: Routledge & Kegan Paul, 1985.

Levinson, Paul. *Mind at Large: Knowing in the Technological Age*. Greenwich, Conn.: Jai Press, 1988.

Levy, Steven. "The End of Literature: Multimedia Is Television's Insidious Offspring." *Macworld*, June 1990, pp. 61–64.

————. "In the Realm of the Censor: The Online Service Prodigy Tells Its Users to Shut Up and Shop." *Macworld*, January 1991, pp. 69–74.

Lombardo, Thomas J. *The Reciprocity of Perceiver and Environment: The Evolution of James J. Gibson's Ecological Psychology*. Hillsdale, N.J.: Erlbaum, 1987.

Lyotard, Jean François. *The Postmodern Condition: A Report on Knowledge*. Trans. Geoff Bennington and Brian Massumi. Minneapolis: University of Minnesota Press, 1984.

McAleese, Ray, ed. *Hypertext: Theory into Practice*. Norwood, N.J.: Ablex, 1989.

McAleese, Ray, and Catherine Green, eds. *Hypertext: State of the Art*. Norwood, N.J.: Ablex, 1990.

McCorduck, Pamela. *The Universal Machine: Confessions of a Technological Optimist*. New York: McGraw-Hill, 1985.

McLuhan, H. Marshall. *The Gutenberg Galaxy: The Making of Typographic Man*. Toronto: University of Toronto Press, 1962.

————. *Understanding Media: The Extensions of Man.* New York: McGraw-Hill, 1964.

McLuhan, H. Marshall, and Eric McLuhan. *Laws of Media: The New Science.* Toronto: University of Toronto Press, 1988.

Mitterer, John, Gary Oland, and J. S. Schankula. *Hypermedia Bibliography.* Technical Report CS-88-02. St. Catherines, Ont.: Brock University, Department of Computer Science, 1988.

Morrison, Jim. *The Lords and the New Creatures.* New York: Simon and Schuster, 1969.

Nelson, Theodore Holm. "As We Will Think." In *Proceedings Online 72: International Conference on Online Interactive Computing.* Uxbridge, Eng.: Online Computer Systems, 1973.

————. *Computer Lib/Dream Machines.* Redmond, Wash.: Tempus Books, 1987.

————. "Computopia Now!" In *Digital Deli*, ed. Steve Ditlea. San Francisco: Workman, 1984.

————. "A Conceptual Framework for Man–Machine Everything." In *Proceedings AFIPS National Computer Conference, June 4–8, 1973.* Montvale, N.J.: AFIPS Press, 1973.

————. "Getting It Out of Our System." In *Critique of Information Retrieval*, ed. G. Schechter. Washington, D.C.: Thompson Books, 1967.

————. *Literary Machines.* Sausalito, Calif.: Mindful, 1990.

Nielsen, Jakob. *Hypertext and Hypermedia.* San Diego: Academic Press, 1990.

Nora, Simon, and Alain Minc. *The Computerization of Society.* Cambridge, Mass.: MIT Press, 1981.

Norton, Roger Cecil. *Hermann Hesse's Futuristic Idealism: The Glass Bead Game and Its Predecessors.* Frankfurt am Main: Peter Lang, 1973.

Ong, Walter J. *Interfaces of the Word: Studies in the Evolution of Consciousness and Culture.* Ithaca, N.Y.: Cornell University Press, 1977.

————. *Orality and Literacy: The Technologizing of the Word.* New York: Methuen, 1982.

Pavic, Milorad. *The Dictionary of the Khazars: A Lexicon Novel in 10,000 Words.* New York: Knopf, 1988.

Pimental, Ken, and Kevin Teixeira. *Virtual Reality: Through the New Looking Glass.* Blue Ridge Summit, Pa.: Windcrest, 1992.

Poster, Mark. *The Mode of Information.* Chicago: University of Chicago Press, 1990.

Rheingold, Howard. *Tools for Thought: The People and Ideas Behind the Next Computer Revolution.* New York: Simon and Schuster, 1985.

————. *Virtual Reality.* New York: Summit Books, 1991.

Romanyshyn, Robert Donald. *Technology as Symptom and Dream*. New York: Routledge & Kegan Paul, 1989.

Rorty, Richard. *Philosophy and the Mirror of Nature*. Princeton, N.J.: Princeton University Press, 1980.

Rosenstock-Huessy, Eugen. *Out of Revolution: Autobiography of Western Man*. Norwich, Vt.: Argo Books, 1970.

Shneiderman, Ben. *Designing the User Interface*. Reading, Mass.: Addison-Wesley, 1987.

Siu, Ralph Gun Hoy. *Ch'i: A Neo-Taoist Approach to Life*. Cambridge, Mass.: MIT Press, 1974.

Slatin, John. "Composing Hypertext." In *The Hypertext/Hypermedia Handbook*, ed. Emily Berk and Joseph Devlin. New York: McGraw-Hill, 1991.

———. "Reading Hypertext: Order and Coherence in a New Medium." *College English* 52 (1990): 870–83.

Spring, Michael B. "Informating with Virtual Reality." In *Virtual Reality: Theory, Practice, and Promise*, ed. Sandra K. Helsel and Judith Paris Roth. Westport, Conn.: Meckler, 1991.

Stoll, Clifford. *The Cuckoo's Egg: Tracking a Spy Through the Maze of Computer Espionage*. Garden City, N.Y.: Doubleday, 1989.

Sutherland, Ivan E. "A Head-Mounted Three-Dimensional Display." In *AFIPS Conference Proceedings*, 1968 Fall Joint Computer Conference, vol. 33. Washington, D.C.: Thompson, 1968.

Tanner, Michael. "The Total Work of Art." In *The Wagner Companion*, ed. Peter Burbidge. New York: Cambridge University Press, 1979.

Traub, David C. "Simulated World as Classroom: The Potential for Designed Learning Within Virtual Environments." In *Virtual Reality: Theory, Practice, and Promise*, ed. Sandra K. Helsel and Judith Paris Roth. Westport, Conn.: Meckler, 1991.

Tufte, E. *The Visual Display of Quantitative Information*. Cheshire, Conn.: Graphics Press, 1984.

Turkle, Sherry. *The Second Self: Computers and the Human Spirit*. New York: Simon and Schuster, 1984.

U.S. Congress. Senate. Committee on Commerce, Science, and Transportation. Subcommittee on Science, Technology, and Space. *New Developments in Computer Technology: Virtual Reality*. Washington, D.C.: Government Printing Office, 1992.

Virilio, Paul. *War and Cinema: The Logistics of Perception*. Trans. Patrick Camiller. New York: Verso, 1989.

Walser, Randal. "The Emerging Technology of Cyberspace." In *Virtual Reality: Theory, Practice, and Promise*, ed. Sandra K. Helsel and Judith Paris Roth. Westport, Conn.: Meckler, 1991.

Weizenbaum, Joseph. *Computer Power and Human Reason: From Judgment to Calculation*. San Francisco: Freeman, 1976.

Whorf, Benjamin. *Language, Thought and Reality*. Cambridge, Mass.: MIT Press, 1956.

Winner, Langdon. *Autonomous Technology*. Cambridge, Mass.: MIT Press, 1977.

Woolley, Benjamin. *Virtual Worlds: A Journey in Hype and Hyper-reality*. Cambridge, Mass.: Blackwell, 1992.

Wurman, Richard Saul. *Information Anxiety*. New York: Bantam Books, 1989.

Yankelovich, Nicole, and Paul Kahn. *Hypermedia Bibliography*. Providence, R.I.: Institute for Research in Information and Scholarship, 1989.

Zuboff, Shoshana. *In the Age of the Smart Machine: The Future of Work and Power*. New York: Basic Books, 1988.

INDEX

The numbers in italics refer to pages in the vocabulary of useful terms.

DATE DUE